G000078944

Indigeneity and the Sacred

Studies in Environmental Anthropology and Ethnobiology

General Editor: **Roy Ellen**, FBA

Professor of Anthropology, University of Kent at Canterbury

Interest in environmental anthropology has grown steadily in recent years, reflecting national and international concern about the environment and developing research priorities. This major new international series, which continues a series first published by Harwood and Routledge, is a vehicle for publishing up-to-date monographs and edited works on particular issues, themes, places, or peoples that focus on the interrelationship between society, culture, and environment. Relevant areas include human ecology, the perception and representation of the environment, ethnoecological knowledge, the human dimension of biodiversity conservation, and the ethnography of environmental problems. While the underlying ethos of the series will be anthropological, the approach is interdisciplinary.

Volume 1
The Logic of Environmentalism: Anthropology, Ecology and Postcoloniality
Vassos Argyrou

Volume 2
Conversations on the Beach: Fishermen's Knowledge, Metaphor and Environmental Change in South India
Götz Hoeppe

Volume 3
Green Encounters: Shaping and Contesting Environmentalism in Rural Costa Rica
Luis A. Vivanco

Volume 4
Local Science vs. Global Science: Approaches to Indigenous Knowledge in International Development
Edited by Paul Sillitoe

Volume 5
Sustainability and Communities of Place
Edited by Carl A. Maida

Volume 6
Modern Crises and Traditional Strategies: Local Ecological Knowledge in Island Southeast Asia
Edited by Roy Ellen

Volume 7
Traveling Cultures and Plants: The Ethnobiology and Ethnopharmacy of Human Migrations
Edited by Andrea Pieroni and Ina Vandebroek

Volume 8
Fishers And Scientists In Modern Turkey: The Management of Natural Resources, Knowledge and Identity on the Eastern Black Sea Coast
Ståle Knudsen

Volume 9
Landscape Ethnoecology: Concepts of Biotic and Physical Space
Edited by Leslie Main Johnson and Eugene Hunn

Volume 10
Landscape, Process and Power: Re-Evaluating Traditional Environmental Knowledge
Edited by Serena Heckler

Volume 11
Mobility and Migration In Indigenous Amazonia: Contemporary Ethnoecological Perspectives
Edited by Miguel N. Alexiades

Volume 12
Unveiling the Whale: Discourses on Whales and Whaling
Arne Kalland

Volume 13
Virtualism, Governance and Practice: Vision and Execution in Environmental Conservation
Edited by James G. Carrier and Paige West

Volume 14
Ethnobotany in the New Europe: People, Health and Wild Plant Resources
Edited by Manuel Pardo-de-Santayana, Andrea Pieroni and Rajindra K. Puri

Volume 15
Urban Pollution: Cultural Meanings, Social Practices
Edited by Eveline Dürr and Rivke Jaffe

Volume 16
Weathering the World: Recovery in the Wake of the Tsunami in a Tamil Fishing Village
Frida Hastrup

Volume 17
Environmental Anthropology Engaging Ecotopia: Bioregionalism, Permaculture, and Ecovillages
Edited by Joshua Lockyer and James R. Veteto

Volume 18
Things Fall Apart? The Political Ecology of Forest Governance in Southern Nigeria
Pauline von Hellermann

Volume 19
Sustainable Development: An Appraisal Focusing on the Gulf Region
Edited by Paul Sillitoe

Volume 20
Beyond the Lens of Conservation: Malagasy and Swiss Imaginations of One Another
Eva Keller

Volume 21
Trees, Knots, and Outriggers: Environmental Knowledge in the Northeast Kula Ring
Frederick H. Damon

Volume 22
Indigeneity and the Sacred: Indigenous Revival and the Conservation of Sacred Natural Sites in the Americas
Edited by Fausto Sarmiento and Sarah Hitchner

Indigeneity and the Sacred

Indigenous Revival and the Conservation of Sacred Natural Sites in the Americas

Edited by Fausto Sarmiento and Sarah Hitchner

First edition published in 2017 by

Berghahn Books

www.berghahnbooks.com

Library of Congress Cataloging-in-Publication Data

A catalogue record for this book is available from the Library of Congress

British Library Cataloguing in Publication Data

A catalogue record for this book is available from the British Library

ISBN 978-1-78533-396-5 (hardback)
E-ISBN 978-1-78533-397-2 (ebook)

Contents

List of Figures and Tables

Figures

Tables

Dedication

We dedicate this volume to the local and indigenous groups in the Americas who are actively working to protect both tangible and intangible cultural heritage and the landscapes in which these are located and enacted, in ways big and small and seen and unseen and to the conservationists and policymakers who are bridging science and humanities in an effort to effectively integrate religion, history, and the political construction of indigenous identities into a driving force of sustainable development throughout the region.

Acknowledgments

We would first like to thank Betty Jean Craige at the University of Georgia (UGA) for support provided by the Willson Center for Humanities and Arts to initiate a faculty discussion group on the notion of indigenous revival and sacred sites conservation, in which the coeditors began their collaboration. We anticipate that this collaboration will extend beyond the publication of this volume and lead to a collaborative research proposal and a long-term emphasis on biocultural heritage in the region.

Support for this initial collaboration, and subsequent meetings that emerged from it, came from several sources within UGA, but several key extramural sources supplemented intramural funding. We thank Mr. Barton Rice, a long-time donor to UGA's Geography Department, for his seed funding. We would also like to thank Mr. Andrew Leighton of the Monterey Institute of International Studies, who facilitated contact with the Andean Alliance for Sustainable Development's artisans of the Sacred Valley of the Inka in the Peruvian village of Písac and helped to distribute handmade *watukuna* to collaborators at one key meeting in 2012. We would also like to extend our appreciation to Ms. Jessica Brown, executive director of the New England BioLabs Foundation and chair of the Protected Landscape Specialist Group of the International Union for Conservation of Nature (IUCN), whose key support helped to bring delegates from Africa, Europe, and Latin America to meet collaborators in the United States.

In-house support was provided by a contribution from the UGA President's Venture Fund through the generous gifts of the University of Georgia Partners and other donors, the Provost's State of the Art Conference Fund, the Willson Center for Humanities and Arts, and the cosponsorship and in-kind support of the Latin American and Caribbean Studies Institute (LACSI), the Institute of Native American Studies (INAS), the Center for Integrative Conservation Research (CICR), and the Geography Department. We wish to give special mention to Michael Adams, Jere Morehead, Nancy McDuff, Nicholas Allen, Richard Gordon, Jace Weaver, Peter Brosius, Thomas Mote, Leara Rhodes, Julie Velásquez Runk, John Schelhas, James Reap, and others for support of this endeavor and for contributing their expertise.

We are grateful for the help of a team of energetic and enthusiast graduate students of the Neotropical Montology Collaboratory, including Luis Candelario, David Charles, Brandon Comb, Ricardo Vercoe, Brandy Gordon, and Alana Shaw, who have assisted with the logistics of in-person collaborations among the authors of this volume. We would also like to thank UGA students and staff members Sarah Burr, Robert Phares, Emily Coffee, and Loretta Scott for their assistance as well. We also thank Barbara Duncan, the Education Director of the Museum of the Cherokee Indian in North Carolina, as well as Ben Steere, for facilitating and guiding a visit of the authors to the sacred landscapes of the Cherokee.

We also gratefully acknowledge additional funding for manuscript preparation, specifically the task of indexing, from Willson Center for Humanities and Arts at UGA, the UGA Geography Department, the Institute of Native American Studies at UGA, and the Latin American and Caribbean Studies Institute at UGA. In addition, the contents of this book were developed under grant #P015A140046 from the U.S. Department of Education.[1] To all, we offer our deeply felt gratitude.

Notes

1. However, those contents do not necessarily represent the policy of the U.S. Department of Education, and the Federal Government does not necessarily endorse them.

Preface

Fausto Sarmiento and Sarah Hitchner

In recent decades, indigenous leaders and activists have taken on unprecedented agency and prominence, leading movements to protect their lands and rights from hegemonic forces in the form of domestic and foreign governments, powerful corporations, and even oppressive conservation and social welfare initiatives. One example is the Wao-Terero, known as Tagaeri (Waorani-Sabela for "red painted feet") or as Awashiri (Kichwa [also spelled Quechua] for "high ground people")—a bellicose group of the Waorani nation that lives in the heart of Yasuni National Park, located in the upper Amazon region of Ecuador. This group received worldwide recognition in 2007—in absentia—when presented the Prince of Asturias' Bartolomé de las Casas Honorable Mention for their resistance to acculturation and for efforts to protect lands they have historically occupied.

The indigenous political presence is so strong in Ecuador that organized protests have paralyzed the country on several occasions, even to the point of taking down presidents and helping to elect new presidents who were more sensitive and responsive to indigenous issues. Furthermore, in Bolivia, an indigenous *cocalero* leader became the democratically elected president for the first time since the creation of the Republic of Bolivia.

In Colombia, the Resguardos Indígenas (autonomous indigenous administrative jurisdictions) have been granted legal recognition of communal land tenure by the national government; management of these indigenous-owned territories is rooted in the government's policy of administrative decentralization and encourages acceptance of traditional jurisprudence and self-determination of almost 40 percent of its national territory. All other Latin American countries, including Brazil, Chile, Argentina, Venezuela, and many smaller nations in Central America and the Caribbean, have developed ministerial units of general secretariats to provide administrative and legal support for indigenous affairs. Groups that were often marginalized by the central government have now acquired enough clout to make high-profile news in newspapers and key radiocasts.

It is clear that throughout the continents of South and North America, the theme of indigeneity has been reinvigorated and has become prioritized in many sectors, mainly education, agriculture, forestry, mining, and social justice. The word "indigeneity" is used officially in international treaties and conventions such as the United Nations Declaration on the Rights of Indigenous People [UNDRIP] and the International Labour Organization Convention 169 [ILO 169, also known as the Indigenous and Tribal Peoples Convention 1989] and by organizations such as United Nations Working Group on Indigenous People and the World Bank. Importantly, both the United Nations and the World Bank explicitly include in their definitions of indigeneity references to the self-identity of indigenous peoples, not just identity imposed by outsiders.

Strengthened legal rights for indigenous peoples granted by external actors, nationally and regionally, and increased agency and intensified cohesion of ethnic identity by indigenous groups themselves have converged into what Paja Faudree (2013) has termed "indigenous revival." As part of the revivalism movement emphasizing religion and traditional cultural practices around the world (Córdova 2014), the indigenous revival brings about the revitalization of culture, language, traditional knowledge, and customary law of the original people, or *pueblos originarios,* of the Western Hemisphere (Jackson and Warren 2005; Warrior 2014). Indigenous peoples in the Americas, while never totally eradicated, were often left behind during the processes of modernization and relegated to reservations, away from the mainstream of daily governance. Too often, discrimination and lack of opportunities were endemic to areas settled by a majority of indigenous people.

The revivalist movement has found inspiration in the First Nations of Canada and has explored the possibility of funneling resources for working against the status quo, in what is also known as the "awakening" (Gumucio 2002). A plethora of nongovernmental organizations (NGOs) has been created in recent years to advance indigenous issues, such as the United Nations Permanent Forum on Indigenous Issues (UNPFII), the Indigenous Peoples' Biodiversity Network (IPBN), the Indigenous Environmental Network (IEN), and the International Work Group for Indigenous Affairs (IWGIA).

In the United States, the National Museum of the American Indian was finally authorized and built in the Smithsonian complex in Washington, D.C. through a collaborative process involving many Native Americans, who chose how their people would be portrayed (NAMAI 2005). It opened to the public in 2004 and faced some criticism for the way in which "subjective personal narrative is privileged above factual evidence, and the deliberate myth-making of an active national revival trumps scholarship" (Muir 2005). However, others heralded its departure from the type of traditional museum display that relegates living cultures to a specific time and place.

In any case, the United States was one of only four countries (the others being Australia, New Zealand, and Canada) to oppose the United Nations General Assembly's adoption of the United Nations Declaration on the Rights of Indigenous Peoples (UNDRIP) in 2007; however, the United States became a signatory when U.S. president Barak Obama announced the country's endorsement of it in 2010 (Australia became a signatory in 2009, and Canada and New Zealand also signed it in 2010).

Revivalists have started to research the pieces of their history and cultures that have been lost through various political processes (from forced assimilation and language loss to outright theft of land and cultural objects) and to reassemble these pieces, as much as possible, in an attempt to reaffirm their indigenous identity. An international workshop was convened by the International Congress of Americanists in Quito in 1997 on the topic of indigenous identity, in which the conveners explained the fallacy of naming them Indians, Amerindians, natives, aborigines, indigenes, or even indigenous; they called for the use of the term "original people" to avoid any bias and to emphasize the notion of the pertinence and persistence of their biocultural heritage (Cruz 2010). The concept of indigenous revival in Latin America requires scholarly analysis, and little has been done to date to connect the *pueblos originarios* with the trend of globalization (McCormick 2013; Adebayo, Benjamin, and Lundy 2014; Warrior 2014); calls for more work on creation narratives to link ancient places and present communities (Christie 2009) are echoed throughout the region.

On the other hand, an important paradigm shift has occurred in the field of conservation in the last several decades. While conservation efforts by Western practitioners have conventionally been based on the idea of "pristine nature," the views of most conservation practitioners have evolved toward acceptance of the idea of anthropogenic drivers that have led, directly or indirectly, to the current assemblages of floral and faunal biodiversity in cultural landscapes (Adams and McShane 1992; Balée 1992, 2013; Crumley 1994; Cronon 1995). This change of framework requires integrating intangible aspects and incommensurable values that are arguably as valuable as the ecosystems and landscapes in which they are embedded (Berkes 2012). This is a difficult task, though mechanisms have been created to preserve cultural landscapes in a more holistic way; chief among these is sacred sites conservation.

Recognizing the momentum that conservation of sacred natural sites (SNS) has gained in recent years, we proposed the ambitious goal of bringing together the most representative and well-respected scholars on both topics, indigenous revival and sacred natural sites conservation, to meet at the University of Georgia in Athens and reflect on the intricate reality of original people's sustainability scenarios, mainly focusing on how communities in the Americas have shown how revaluing past practices of observance, spirituality, and the veneration of sacred sites can lead to more effective and equitable conservation outcomes.

Much work has taken place in mapping significant cultural sites and landscape modifications, as well as other changes in land use and land cover, in indigenous territories. Hitchner (2009a, 2009b, 2013 and Hitchner et al. 2010) researched the Kelabit Highlands of Sarawak, Malaysia and engaged in community-led efforts to map cultural sites including megaliths, old longhouse sites, burial sites, and other sites of historic or mythological significance. This research was based on a preliminary visit to the area in 2005, when local people expressed the need to map cultural sites as part of larger community efforts to preserve cultural traditions and the local language for future generations of Kelabit, who are becoming increasingly urbanized.

This research was conducted in a state in which indigenous land rights are a very politically sensitive topic, with a geographically dispersed community whose members have differing levels of power to make decisions about their homeland, and in a landscape that is highly contested and for which many plans are being made by many different internal and external actors. The situation is no different in Ecuador, Costa Rica, and Argentina, as Sarmiento (2010, 2013) has demonstrated. But our goal with this volume is to bring attention to the larger geographical problematic: not only the graphic representation of the conflictive occupation of the land but also the plight of rural areas or farmscapes of the Americas that are being transformed—physically as well as culturally and spiritually. As noted, there has been a growing emphasis on incorporating land use history and current land use practices into conservation planning (Sarmiento, Russo, and Gordon 2013). Amid increasing recognition that exclusionary conservation is unjust and ineffective and that flawed models of "pristine wilderness" do not recognize landscape modifications or agroecological diversity created or enhanced by the original people, resulting in biocultural heritage, the need to document and design appropriate frameworks requires geographical enquiry.

However, many conservation projects only superficially include indigenous and local community members, and they often regard them merely as holders of traditional ecological knowledge (TEK) or creation myths and not as complex agents with rights to assess and analyze the pasts and guide the futures of the cultural landscapes in which their cultures are embedded. Furthermore, because SNS are key elements of biocultural heritage conservation, they also play an important role in global environmental change adaptation strategies in the hemisphere (Sarmiento and Viteri 2015). As discussed in more detail in the conclusion, we fully acknowledge the problematization of the notion of SNS, sacred landscapes and sacred landscape features such as mountains and forests. Dove, Sajise, and Doolittle (2011: 7) provide the insight that "the notion of sacred forests is most often used today not by anthropologists or other scholars but by community and environmental activists, and this usage is typically quite normative and uncritical in nature."

The aim of this volume is to show the links between indigenous identity and cultural revival in the context of SNS and landscapes, whether used as a political tool for increased self-governance or revered as places where gods literally dwell. This volume collects a multiplicity of voices, including an indigenous shaman, a Jesuit priest, employees of the U.S. National Park Service, several Latin American scholars, and others, that present different manifestations of SNS, both physically and ideologically. Some are more critical of the categorization of certain places as SNS than others, but all recognize the role that it can play in the maintenance and revitalization of indigenous identity.

It was with these important theoretical and pragmatic streams that the recently created Neotropical Montology Collaboratory (NMC) at the University of Georgia planned, organized, funded, and successfully convened an international conference, as many of the most important sacred sites for the original people of the Americas are located on mountains and in mountainous landscapes. Through the work of NMC, several research clusters were engaged in the conversations prior to establishing the event. The NMC obtained a grant from the Willson Center for Humanities and Arts at the University of Georgia to develop a discussion group on campus to tackle the idea and bring it to fruition. Coeditors Sarah Hitchner and Fausto Sarmiento met as part of this initial effort.

Other links secured institutional cosponsorship and in-kind support from the Geography Department, the Center for Integrative Conservation Research (CICR), the Institute of Native American Studies (INAS), the Latin American and Caribbean Studies Institute (LACSI), the Tate Student Center, and Bolton Hall's Food Services. The Geography Department's generous donor, Mr. Barton Rice, provided the seed money for a miniconference on the topic. Other campus units cosponsored the idea and were instrumental in securing intramural funding from the President's Venture Fund through the generous gifts of the University of Georgia Partners and other donors, the Provost State of the Art Conference fund, and the Willson Center for Humanities and Arts. Extramural funding came from the Exposition Foundation of Atlanta, Inc. and the New England BioLab Foundation.

The international conference "Indigenous Revival and Sacred Sites Conservation" was convened at the University of Georgia in Athens on 5–7 April 2012. In the following year, selected participants were contacted, and the plan for the edited volume materialized. The final submission of this volume includes key scholars to complement the book; so it is not only a book of conference proceedings but a scholarly effort to reflect the current state of knowledge on the topics of indigenous revival and sacred sites conservation in the Americas. We are also grateful to have heard the perspectives of our colleagues from Ghana in Africa and from France, Switzerland, and the Netherlands in Europe, whose work may not be included in this volume but whose participation in the conference was nonetheless deeply appreciated by all attendees.

Organizationally, we divided the book into three parts. In the first part, we provide an overview with a prologue written as a direct account from a shaman of the Cofán Nation in Ecuador, hence presenting anecdotal, inspiring, and thought-provoking ideas with a different voice and tone than those of the following chapters of the book, which were written by scholars. In Chapter 1, conservation management scenarios of sacred natural sites are presented by two leaders in the field who emphasize the importance of SNS in efforts to protect the cultural dynamics and natural diversity necessary for a sustainable future. In Chapter 2, the faithscape of the Americas is explained by a distinguished mountain geographer who frames the spirituality of the region within the larger context of structural change, noting that intricate processes of endogenous production, as well as hegemonic impositions of exogenous replication, are at play. Both chapters offer the reader not only an explanation of the past and a description of the present but also a perspective into the future of sacred sites and indigeneity in the Americas.

In the second part, we frame the sacred sites in the context of indigenous mindscapes throughout the region. In Chapter 3, two leaders of Andean sustainability explain the binary perspective of the *Buen Vivir* as a guiding principle for Andean conservation and development. In Chapter 4, issues regarding the revival and sustenance of indigenous people are presented by a world leader on sacred mountains, who incorporates an ethnographic component into reading the landscape of the Americas with dual perspectives of mystic and pragmatic understanding. In Chapter 5, we explore the notion of the construction of Andean identity associated with the ritualized observance of frozen mummies in the high Andes mountains with the insights of the only female high-altitude archaeologist working on the sacred peaks of the Andes.

In Chapter 6, we include the views of a renowned Latin Americanist, who helps to identify the changing images and dimensions of indigenous identity in space and time, mainly with his long-term effort to define what *lo andino* really means. Finally, in Chapter 7, we include the analysis of two leading experts on governmental efforts to elucidate the cultural and spiritual values embedded in places of significance that are protected by the National Park Service of the United States, including many SNS.

In the third part of the book, we include case studies from foremost researchers to illustrate specific trends in the region. In Chapter 8, collaborative archaeology is described as a tool to preserve SNS in the mountains of the U.S. state North Carolina. In Chapter 9, we include an ethnoecological analysis of the biocultural sacred sites by two active scholars researching Mexico's rich ethnic landscape. In Chapter 10, a conservation biologist explains the new dimensions of territorial conservation amid the biocultural heritage paradigm of politics and management in Ecuador.

Finally, in Chapter 11, we enlist the views of a renowned ethnobotanist and Jesuit priest, who emphasizes the need to recuperate traditional ethnobotanical knowledge for obtaining sustainable development in the selva region of the Peruvian Amazon. We recognize that these chapters are not fully representative of the regions of the Americas; notable areas and peoples are missing, such as the Arctic peoples of North America and the First Nations in Canada. However, we have tried to be as inclusive as possible and would like to expand our regional analysis in further work.

A field trip to the mountains of Georgia and North Carolina culminated with a visit to the Chattahoochee National Forest, the Nikwasi Mound sacred site, and the Museum of the Cherokee Indian, whose education director, Dr. Barbara Duncan, recited the compelling words of John Standingdeer, an elder from Cherokee, North Carolina. He is a leader in the revival of his culture as well as in the maintenance of the sacred cultural landscapes of the southern Appalachians, and his words will reverberate in the minds of the participants. We introduce them here so that readers of this book can also reflect on his message:

> Did they understand, that the mountains are everything to us? They gave us everything. The Sioux had the buffalo and almost made them their god. The Navaho had the sheep from the Spanish, and that gave them everything. They didn't make them their god. We didn't make it that way, but the mountains gave us everything. Everything we needed.
>
> And when our people had to leave here, we were exiled. Some people committed suicide rather than leave. Do they know what it means that we held on to this small part of the mountains?
>
> My dad said, the mountains will always protect us. No matter what happens. And you know the storms this past year, hit all around us? But here we were protected.

> My dad, if you could see him, he loves to be out in the woods, just to be there. And the way he moves through the woods, he belongs here. We belong here. Yes, people can go to the other towns, around here; and they know the white people don't like them. But here, we're at home.
>
> Did these people understand that we have held on to this? What the mountains mean to us?
>
> When I went away from here, I was eleven, we moved to Oklahoma to help my sister's breathing. I hated it. Cherokee people there sang the songs so slow, and so sad. The water didn't taste right, and the air didn't smell right. Nothing was right. When we came back to the mountains, it was wonderful.
>
> When my mother was dying, she said she dreamed that her father and Uncle Morgan brought her a cool drink of water from the branch, and it tasted so good. People don't know what that means, to drink from the stream, and it's cold, and it tastes so good. No other water tastes like that.
>
> The way we used to play, just out in the woods. We barely had toys; we'd just go play outside. I remember once an old Indian man—probably my age now [laughs]—he came back from the war. He'd been at the invasion of Normandy. He took my brother and me up to Mingo Falls when there was no trail there. It was so steep, my brother started leaning backwards. He could have fallen off the mountain. And this man, he'd reach over and grab my brother's collar, stand him back up.
>
> Where we played and what we did—swinging on grapevines and they'd break—it's a wonder any of us lived through it. But we did. And this is our home. It gives us everything. Do they understand that? (Standingdeer 2012)

As editors, we do not claim to understand. But we have attempted to listen. This volume reflects our best efforts to capture what is so difficult to put into words and to assist each of our contributors in expressing their experiences, ideas, convictions, and different ways of knowing with as much clarity as possible. Any mistakes or shortcomings are entirely our own.

Fausto O. Sarmiento, Ph.D., professor of Geography; director of the Neotropical Montology Collaboratory, University of Georgia; expert in Andean cultural landscape conservation. Sarmiento was chair of the American Association of Geographers' Mountain Geography Specialty Group and the International Research and Scholarly Exchange Committee. He taught as visiting professor in Costa Rica, Spain, Argentina, Chile, and Ecuador. He was awarded a plaque from the Centro Panamericano de Estudios e Investigaciones Geográficos ([CEPEIGE], the PanAmerican Center for Geographic Research and Studies). He is the author of *Montañas del Mundo: Una Prioridad Global con Perspectivas Latinoamericanas.*

Sarah Hitchner, Ph.D., assistant research scientist, Center for Integrative Conservation Research, and adjunct professor of anthropology, University of Georgia, Athens. Hitchner is a cultural anthropologist specializing in sacred sites and cultural landscapes of Southeast Asia.

References

Adams, Jonathan S. and Thomas O. McShane. 1992. *The Myth of Wild Africa: Conservation without Illusion.* Berkeley: University of California Press.

Adebayo, Akanmu G., Jesse J. Benjamin, and Brandon D. Lundy, eds. 2014. *Indigenous Conflict Management Strategies: Global Perspectives.* Plymouth, UK: Lexington Books.

Balée, William. 1992. "People of the Fallow: A Historical Ecology of Foraging in Lowland South America." In *Conservation of Neotropical Forests: Working from Traditional Resource Use,* edited by Kent H. Redford and Christine Padoch, 35–57. New York: Columbia University Press.

Balée, William. 2013. *Cultural Forests of the Amazon: A Historical Ecology of People and Their Landscapes.* Tuscaloosa: University of Alabama Press.

Berkes, Fikret. 2012. *Sacred Ecology.* 3rd ed. New York: Routledge.

Christie, Jessica Joyce. 2009. *Landscape of Origin in the Americas: Creation Narratives Linking Ancient Places and Present Communities.* Tuscaloosa: University of Alabama Press.

Córdova, Fabiola. 2014. "Weaving Indigenous and Western Methods of Conflict Resolution in the Andes." In *Indigenous Conflict Management Strategies: Global Perspectives,* edited by Akanmu G. Adebayo, Jesse J. Benjamin, and Brandon D. Lundy, 15–31. Plymouth, UK: Lexington Books.

Cronon, William. 1995. "The Trouble with Wilderness, or Getting Back to the Wrong Nature." In *Uncommon Ground: Toward Reinventing Nature,* edited by William Cronon, 69–90. New York: Norton.

Crumley, Carole L. 1994. "Historical Ecology: A Multidimensional Ecological Orientation." In *Historical Ecology: Cultural Knowledge and Changing Landscapes,* edited by Carole Crumley, 1–13. Santa Fe: School of American Research Press.

Cruz, Alberto. 2010. *Pueblos Originarios en América: Guía Introductoria de su Situación.* Pamplona: Aldea Alternatiba Desarrollo.

Dove, Michael R., Percy E. Sajise, and Amity A. Doolittle. 2011. "Introduction: Changing Ways of Thinking about the Relations between Society and Environment." In *Beyond the Sacred Forest: Complicating Conservation in Southeast Asia,* edited by Michael R. Dove, Percy E. Sajise, and Amity A. Doolittle, 1–34. Durham: Duke University Press.

Faudree, Paja. 2013. *Singing for the Dead: The Politics of Indigenous Revival in Mexico.* Durham: Duke University Press.

Gumucio, Cristián Parker. 2002. "Religion and the Awakening of Indigenous People in Latin America." *Social Compass* 49, no. 1: 67–81.

Hitchner, Sarah. 2009a. "The Living Kelabit Landscape: Documenting and Preserving Cultural Sites and Landscape Modifications in the Kelabit Highlands of Sarawak, Malaysia." *Sarawak Museum Journal* 66, no. 87: 1–79.

Hitchner, Sarah. 2009b. *Remaking the Landscape: Kelabit Engagements with Conservation and Development.* Doctoral dissertation. University of Georgia.
Hitchner, Sarah. 2010. "Heart of Borneo as a *Jalan Tikus:* Exploring the Links between Indigenous Rights, Extractive and Exploitative Industries, and Conservation at WCC [World Conservation Congress] 2008." *Conservation and Society* 8, no. 4: 320–330.
Hitchner, Sarah. 2013. "Doing High-Tech Collaborative Research in the Middle of Borneo: A Case Study of e-Bario as a Base for the Transfer of GIS Technology in the Kelabit Highlands of Sarawak, Malaysia." *Journal of Community Informatics* 9, no. 1. Available at: http://ci-journal.net/index.php/ciej/article/view/475/976. (Accessed 9 September 2013).
Hitchner, Sarah, Florence L. Apu, Supang Galih, Lian Tarawe, and Ellyas Yesaya. 2009. "Community-Based Transboundary Ecotourism in the Heart of Borneo: A Case Study of the Kelabit Highlands of Malaysia and the Kerayan Highlands of Indonesia." *Journal of Ecotourism* 8, no. 2: 193–213.
Jackson, Jean E. and Kay B. Warren. 2005. "Indigenous Movements in Latin America, 1992–2004: Controversies, Ironies, New Directions." *Annual Review of Anthropology* 34: 549–573.
McCormick, Katie. 2013. *Reviving Indigenous Voices: Ideologies, Narratives and Methods.* B.A. thesis. Swarthmore College. Tricollege Digital Repository.
Muir, Diana. 2005. "National Myth of the American Indian." *Claremont Review,* 4 March 2005. Available at: http://www.freerepublic.com/focus/f-news/1358787/posts (accessed 1 September 2016).
NAMAI. 2005. *The Native Universe and Museums in the Twenty-First Century: The Significance of the National Museum of the American Indian.* Washington, D.C.: Smithsonian Institution.
Sarmiento, Fausto O. 2010. "The Lapwing in Andean Ethnoecology: Proxy for Landscape Transformation." *Geographical Review* 100, no. 2: 229–245.
Sarmiento, Fausto O. 2013. "El Revivir Indígena, los Paisajes Culturales y la Conservación de Sitios Sagrados." *Revista Parques* 1: 1–13.
Sarmiento, Fausto O., Ricardo Russo, and Brandilyn Gordon. 2013. "Tropical Mountains Multifunctionality: Dendritic Appropriation of Rurality or Rhyzomic Community Resilience as Food Security Panacea." In *Multifunctional Agriculture, Ecology and Food Security: International Perspectives,* edited by J. Ram Pillarisetti, Roger Lawrey, and Azman Ahmad, 55–66. New York: Nova Science Publishers.
Sarmiento, Fausto O. and Xavier Viteri O. 2015. "Discursive Heritage: Sustaining Andean Cultural Landscapes Amidst Environmental Change." In *Conserving Cultural Landscapes,* edited by Ken Taylor, Archer St. Clair, and Nora Mitchell, 309–324. New York: Routledge.
Standingdeer, John C. Jr. 2012. Personal communication May 6, 2012.
Warrior, Robert, ed. 2014. *The World of the Indigenous Americas.* New York: Routledge.

Part 1

Geographies of Indigenous Revival and Conservation

Introduction

Whose Sacred Sites?
Indigenous Political Use of Sacred Sites, Mythology, and Religion

Randall Borman
Shaman, Cofán Nation of Ecuador
Executive Director, Cofán Survival Fund/Fundación Sobrevivencia Cofán

Let me begin by talking just a little about the historical context of the present interest in indigenous sacred sites. I won't spend a great deal of time here and will make no big attempts to create a bibliography of references—most of what I am saying is well known to indigenous peoples and those who are interested in them. However, I think it is important that we establish at the onset the origin of the concept of "sacred sites" as we understand it today.

The Western world has a tremendous respect for what it terms "sacred sites." Within its own Judeo–Christian tradition, sacred sites include churches and cathedrals; locations such as the Wailing Wall, the Mount of Olives, and others where crucial religious events have been played out; and lesser sites where miracles of one sort or another have been purported to have taken place. Graveyards are frequently considered sacred sites, and reverence is often attached to the birthplace or home of a particularly "holy" writer, preacher, or exemplary personage. And, as the Western world has increasingly secularized, these same attitudes and respect have been transmitted to icons viewed as ideals within political and economic systems: the Lincoln Memorial, the World Trade Center, and others.

However, sacred sites of other peoples were almost universally seen as threatening and dangerous by the Western world during its early expansion. Spanish conquistadores, English pioneers, French priests, and Dutch merchants all sought to subdue and appropriate any and all sacred sites they met in their travels as a means of spreading Christianity and anchoring their cultural hold on their "conquests." It was not until the middle of the twentieth century, in the wake of the ultimate intolerance, the Holocaust, and deeply aware of its own guilt in having effectively done the same thing with indigenous peoples of both Africa and the Americas, that the Western world began to create a framework of respect for the sacred sites of other people.

As usual with the Western world, once the idea got going, there was no stopping it. The U.S.-based counterculture of the 1960s not only espoused an idealized and stereotyped vision of the Native American world but created a whole series of sacred sites around dimly understood remnants of cultural lore. Latin American countries, while still doing their best to absorb or eliminate their own indigenous peoples, moved to create revisionist histories in which the Spanish conquistadores were the villains and the noble Aztecs, Incas, and others were helpless victims who came from a far purer form of life. Sacred sites not already preempted by European religion now became symbols of awe among a generation of young people disillusioned with their own Judeo–Christian cultural background who were searching for relevance in other spiritual worldviews.

By the late twentieth century, this attitude shift had increasingly invaded the legal and political world. The last decade of the century was named "the Decade of Indigenous Peoples." The International Labor Treaty of 1996 had an entire chapter dedicated to indigenous rights. Most American countries established legal structures to protect their indigenous peoples, and numerous laws and regulations were passed to protect sacred sites. Given this favorable legal and political climate, indigenous peoples worldwide began to organize and work to rebuild damage done during centuries of repression.

Much of this work was aimed at regaining land areas and specific cultural rights for access to resources. Some of it was purely political, and not all of this was positive. In many cases, Western pseudointellectuals actively moved in to preempt and try to direct processes to fit their idealized vision of "indigenousness." Especially in South America, big business interests moved to infiltrate and take over the "indigenous movements" for their own ends. Meanwhile, the temporary lifting of pressure on indigenous cultures allowed many groups to consolidate their positions and enter the twenty-first century in better shape than ever before.

Thus, by the first years of the twenty-first century, indigenous groups around the world had progressed dramatically in gaining a voice and in our ability to defend our rights. However, collectively, we were very aware that the mechanisms we were using had been created by the Western world and that for us to interact effectively we needed to understand very clearly the Western concepts behind these mechanisms. At least at our level in the Ecuadorian Amazon, we quickly became aware that we were still dealing with a deck seriously stacked against us.

To understand our problem, let us look briefly at what indigenous culture is all about—for that matter, what any culture is all about. Culture can be defined as a particular group of people's relation to its physical, spiritual, and social environments. A quick and easy way to assess a culture's priorities is to look at the language—if culture is a people's relation to its physical, spiritual, and social environments, language is the description of that relationship. The minute we look at intact languages, we realize that even where social and spiritual manifestations of any culture seem to define that culture's existence, the relation with the physical environment remains key. Without our physical environment, our social and spiritual environments have no place to develop.

This has led to indigenous groups' tremendous emphasis, worldwide, on the importance of recovering territories. Here we are not talking about just hunting or fishing rights or payoffs for exploitation of resources: we are talking about the tremendous importance to our cultures of being able to regain control over our territories, if we are to continue to maintain our cultures.

We quickly became aware that this was extremely threatening to the Western world. An amusing moment for me was when I was talking with the principal of my children's school. He was from Michigan and in any circle would be understood as a sympathetic, idealistic, well-educated, and well-balanced member of the Western world. He would be the first to defend human rights and shares with many white Americans the burden of guilt for his culture's treatment of indigenous peoples in the nineteenth century.

However, some twist of the conversation brought up the subject of property in Michigan, and he suddenly became a different person. It appeared that a band of Native Americans, flush with money from casinos established on tribal lands, were aggressively going out to buy back their traditional lands. To the horror of my friend, they were going up to farms and knocking on doors and offering cash on the barrelhead for properties. I asked, mildly, where the problem was—if they were offering fair value, in cash, to farmers who wished to sell, wasn't that the proper American way to do business? But somehow, the potential resurgence of indigenous control over vast areas of Michigan farmlands was extremely threatening to him, and somehow very un-American.

The Western world's response to this threat is interesting in its complexity. One of the most common methodologies is to try to divert interest from real issues. In our case (that of the Cofán Nation of northwest

Amazonia), we have had a number of nongovernmental organizations (NGOs) show up to try to help us "rescue" our religion. Where are your ceremonies? Why don't you have shamans? What are your sacred feasts? Why don't you dress the way people did a generation ago? The message is clear. If we can justify helping you to "regain" a redefined and nonthreatening "culture," we will no longer have to deal with an uppity bunch of Indians who are gaining control of lands we want for our objectives.

Even the government has gotten into the act: "Let us help you develop musical groups, and please get together a dance troupe dressed in 'traditional' clothing. But please don't talk to us about keeping your streams pure and your forests intact. We need your resources far too badly. Meanwhile, we will help you preserve your 'heritage,' just so long as you don't keep us out of your lands."

I think I can speak for most indigenous peoples when I say that our cultures are generally very pragmatic. We associate deeply with our physical, social, and spiritual environments, but the association is nothing if not practical. While we often express ourselves in deeply idealistic terms, the daily lives of our people are eminently down-to-earth. This is the way all humans have interacted with their environments since the beginning of time. This does not mean that the sacred does not exist for us. But it seldom mirrors the Western ideal.

It is within this context that we seek to understand the Western world's concept of "sacred sites." At least among my people, we have no temples, nor do we have centralized locations for "worship." Our spiritual interaction with our environment is deeply interwoven with our physical interaction. Much of our spiritual expression is aimed at providing our people with purely physical benefits—safety, food, health, and an environment in which our social expression is maximized. For this, we need our lands above all. In a very real sense, our most sacred site is our land. But this doesn't help us as we try to interface with an aggressive "Other" in the form of the Western world. So we need to muster every possible argument and every possible legality to be able to come to grips with this Other, and to do so means using the Western concepts—even when they are not accurate—to our advantage. This means using Western idealism, Western legal processes, Western mysticism, and Western preconceptions for our ultimate goal—that of recovering and gaining control of our territories.

In our specific case, this was the situation we faced as we came to understand the importance of sacred sites in the Western understanding of indigenous cultures. We have no word for sacred site. Our graveyards are places of sadness and distance; we look for convenience in burying our dead (soft earth), a place far enough away from living humans so that possible inimical spirits released by the death of a person of power will not molest the living, and a place not easily visible so that memories will be eased and not awakened constantly. But sacred in the sense of being hallowed or something we wish to preserve at all costs?

No. The river is constantly changing course, our people move our villages, the forest or the river reclaim locations, and the future generations do not feel obliged to care for the sites of their ancestors in the face of natural processes. We watched with a little sadness, but more amusement, as the burial site of my village's chief, a noted shaman who died in the mid-1960s, was claimed by colonists and eventually became the site of a bar/dance hall. We figured if anything was left of his spirit tied to the spot, he probably thought it was great to have constant parties over his bones.

The same is true of other "important" sites. The huge rock stump of the Fish Tree, which figures deeply in our creation myth, was easily seen until a couple of generations ago, when the river it is located on began to change course. Now the stump is partially hidden under brush and alluvial rocks. It is a point of interest to the younger generation, but no special awe surrounds it. Our one active volcano in our territories has long been known as a site of power and has been frequented by generations of shamans seeking spiritual strength. However, no mass pilgrimages or prayerful attitudes accompany our knowledge. We recognize the mountain's power, both physical and spiritual, but no one was especially upset when the Ministry of Environment claimed it as part of a park. The list could go on. The bottom line is that humans are living beings moving through a changeable and dynamic series of environments, where the idea of trying to hold to a particular site in a special way is a luxury with little deep meaning.

However, as we became aware of Western reverence for sacred sites, we immediately recognized the potential for using this Western preconception to our advantage. Discussion within the communities began in the late 1980s and continues today, as people wrestle with the concept and its implications as we seek to protect our territories and, in a very real sense, create sacred sites.

I have mentioned graveyards. One of our first attempts at using Western concepts to recover our territorial rights was to stop a road, being built to access gravel, from crossing a quiet forest we had designated as an area for burials. If we had merely opposed the road on the grounds that the forest is a productive and necessary part of our lands, we would have been laughed at. However, when we presented it as a graveyard, the reaction was "Gosh, we're sorry ... thanks for telling us, we'll figure out a different route."

The Fish Tree stump and our tame volcano El Reventador were obvious fits, and both now are within Cofán-managed territories. Likewise, a region in the Cofán Bermejo Ecological Reserve remains important as an isolated area where true shamans can still seek power without the constant static from both outsiders and Cofáns who no longer follow the necessary taboos during power searches.

Given the power of the model, we then proceeded to invent sites—most notably the Falls of the Rio Coca. This huge and impressive falls has long been part of Cofán territory, and while no one especially worshipped it, it has certainly been the source of much comment and awe within the Cofán Nation. Thus it was no stretch to declare this a sacred site in spite of its lack of direct spiritual implications. In this, we were imitating the secular sacred sites that have become so common in post-Christian Western culture.

These are small examples of a phenomenon I have seen happening repeatedly within the indigenous world. I was recently with friends from a small Northwest tribe. They showed me the pit from which their first people came and mentioned the importance of their graveyards in their fight for territory. Another friend from the Great Plains area described the tremendous battle over sacred sites in the Black Hills. Yet another friend described the importance of a "sacred" waterfall and salt lick in the middle of Peru.

Thus, the figure of the sacred site has become an important political tool in the hand of modern indigenous activists as we seek to regain control over our territories. We are not lying when we call these sites sacred. But the meaning of sacred for us is very different than for the Western world. The bottom line for us is that all of our territory is sacred in the sense that it is deeply, powerfully imbued with spiritual reality. To take any particular location and call it more sacred makes little sense within a worldview in which our interactions are with all of our environments, and all are sacred. But if the singling out of particular locations will aid us in dealing with the Western world as we seek to maintain our culture, we can meet them halfway.

There are two final comments I would like to make concerning this.

One is that, as we work to fit into an ideal imposed upon us by the outside world, we create a new reality. While we of this generation recognize that much of how we present ourselves to the outside world is in response to its demands rather than our own realities, the younger generations will grow up in a world where Reventador is a seriously sacred location, where the Fish Tree stump is an important monument, and where our often tongue-in-cheek rhetoric defending the Coca Falls is truly the description of an age-old relationship. It is sad that this has to be. It has already happened in many corners of the indigenous world, and perhaps it is revisionism at least as strong as the Latin American anticonquistador movement of the 1960s and 1970s.

Finally, I would like to end with a small story. During the early 1990s, my small community of Cofán was able to recover control over more than 300,000 acres of forests, rivers, swamps, and hills. As chief of the community, I became deeply aware of my stewardship, not only of the people in my territory but of the trees, the animals, the water, the air—we were living in a harmony that included all that was there. However, it was not until one night several years later that the full implications of stewardship came home to me.

There is a small mythical creature in our forests called the *dusese* (doo-say-say). It reputedly looks like a very tiny human, perhaps a meter tall, and travels almost exclusively at night. Usually minding its own business, it is potentially dangerous to humans, and people stay inside and under the covers when one comes near a village. Its call is an ascending whistle, quite similar to our local species of the cuckoo but more intense and different. I say "mythical" advisedly, as there is little room in my Western-trained belief structure for the *dusese*. However, on this particular night, at about nine o'clock, a *dusese* began to call outside of our house. The feel, as it were, was totally different from the cuckoo, and the hair on the back of my neck crawled.

Then I realized that this creature was coming from upriver—where the oil company is working, where colonists are daily expanding their activities, where towns are being built, where the water is polluted by miners and oil production—and an enormous sympathy came over me, and I stood up, went to the window, and spoke, saying, "You are welcome to stay with us, for we, too, understand the need for this forest." The next morning, other community members described the *dusese*'s travels—our farthest upriver households, some fifteen kilometers away from our central village, were the first to hear it, around 7:30. It had passed by each of our households in the area until it got to us, and by 9:30 had continued past our lowest household—no cuckoo would have moved so far or so fast.

Sacredness for us is about the complete environment—an environment rich in spirituality, in importance for our social lives, and for our physical survival. It is the Coca Falls, it is Reventador, it is the Fish Tree stump, it is our graveyards. But it is also the stewardship of rivers, trees, animals, and *duseses*. It is a world where we are part of a whole. To isolate any part and set it up as special may have its uses, but it is the whole that lends even those parts their sanctity, and without the whole, we become nothing more than isolated folk dancers disguised in garments that no longer fit.

Randall Borman is Chief of Territories of the Cofán Nation, headwaters of the Napo River, Ecuador and vice-president of the Cofán community of Zabalo. He is also executive director of the Cofán Survival Fund/Fundación Sobrevivencia Cofán. He is an indigenous rights activist and invited speaker for TEDTalk—Amazonia. He holds an honorary LLD (doctor of laws) degree from Northwest University, Kirkland, Washington.

CHAPTER 1

Connecting Policy and Practice for the Conservation of Sacred Natural Sites

Bas Verschuuren, Robert Wild, and Gerard Verschoor

Introduction: Sacred Natural Sites, Past, Present, Future

Sacred natural sites, pilgrimage routes, and places of heightened spiritual or cultural significance are found in every country and on almost every continent. They form an interconnected fabric that is often neither sufficiently understood nor recognized in conservation policy and management. This partially intangible fabric is part of the everyday reality of many rural and indigenous people who have, often for many generations, played a key role in the governance and management of these sites of worship and their surrounding environments and landscapes. With the goal of maintaining the cultural and biological diversity that sacred natural sites embody, conservationists are tasked with the question of how to recognize and accommodate the needs of sacred natural sites' custodians in conservation management, planning, and policy.

In this chapter, we explore sacred natural sites conceptually and identify opportunities for including them in conservation policy and practice. We suggest some possible ways forward that are cognizant of the ontological diversity that shapes the existence of sacred natural sites.

The resilience of interconnected biological and cultural systems underscores the vitally important role of local and indigenous communities, faith groups, and others that seek spirituality in nature. As many sacred natural sites are increasingly under pressure from uncontrolled development activities, their resilience is often dependent on the role they play—not only in peoples' socioeconomic realities but especially regarding their spiritual well-being (Delgado et al. 2010).

Ultimately, a broad variety of interest groups come together in the conservation of sacred natural sites, and lessons on conservation management and governance can be drawn from these encounters. Many of these lessons are rooted in a variety of worldviews that are influenced by historical, cultural, religious, and spiritual dimensions related to nature (Posey 1999; Berkes 1999) and that encode important ethical and moral concerns and behaviors related to human well-being and the sustainable use of ecosystems. These worldviews are themselves dynamic and adaptive and therefore do not conform to static or prescriptive understanding

Sociologists, cultural anthropologists, and political ecologists traditionally portray dilemmas of resource management and governance in terms of a clash between neocolonial (or neoliberal) mentalities and indigenous ways of being (e.g., Blaser 2009; Büscher et al. 2012; Hunt 2014). This is perhaps not surprising given the historic legacy of human rights abuses by conservationists. Yet the practices and politics of conservation are changing and diversifying. In this chapter we assess some of the emerging spaces in international policy, as well as recent developments in conservation practices, in order to identify opportunities for the conservation of sacred natural sites. We argue that as cultural and spiritual values of nature become increasingly recognized in conservation, explicit attention to the ontological dimension of these values will be key in developing more inclusive forms of biocultural conservation (Mathez-Stiefel 2007; Verschuuren 2010).

Sacred natural sites comprise some of the oldest conserved areas on Earth, yet they have only recently gained attention among conservationists. To acquaint the reader with the notion of sacred natural sites, we first present the conceptualization and growing popularity of sacred natural sites in the modern conservation movement and then describe how a series of conferences and the development of guidelines for protected area managers have worked to sensitize conservationists to sacred natural sites and their custodians. We then reflect on the spaces that have opened up in international policy and the opportunities these offer for the conservation of sacred natural sites and provide two examples of how international legislation could be implemented nationally.

We conclude by making suggestions for the way forward. We posit that the opportunities identified in the policy frameworks discussed in this chapter need to be cautious and also developed with a specific set of sensibilities if and when they are implemented. Importantly, we argue that the conservation, management, and policy of sacred natural sites should follow biocultural conservation approaches that consider the cultural, natural, and spiritual values of sacred natural sites; that is, they should be rights based and work to enhance the ontological self-determination of their custodians.

Sacred to Whom, Where, When?

Sacred sites can consist of manmade structures such as temples, shrines, and pilgrimage roads, but many natural places and specific plant and animal species are also of special sacred significance to certain peoples and play a vital role in their well-being. These distinct places are an expression of a spiritual dimension in which nature is often animated, meaning that human values are attributed to nature and elements of nature. For example, groves of trees can be the homes of ancestors in Africa (Sheridan and Nyamweru 2008), and in northern California, rocks may be imbued with spirits of a deity (Theodoratus and LaPena 1994). In both cases, peoples' worship of groves and rocks involves an interpersonal interaction between humans, nature, and the spirit world that is part of their ontology.

In addition to the sacred sites of indigenous peoples, many sacred natural sites exist that are devoted to holy symbols, saints, and practices of mainstream religions such as Jainism, Judaism, Islam, Buddhism, Christianity, and Hinduism. Many of these sites have lost their specifically local character and often comprise manmade structures such as mosques, chapels, and temples. In some cases, these buildings have been deliberately placed at sites that were previously inhabited by spirit beings as a means to their extirpation or cooptation (Byrne 2010). Many sacred sites of mainstream religions are becoming increasingly urbanized or commercialized at the expense of natural elements, for example where large-scale pilgrimages are being accommodated (Verschuuren and Wild 2012). As cases such as Machu Picchu, Mount Kailash Rapa Nui (Easter Island), and Mount Sinai show, this can become a threat to biodiversity and other conservation values and hence a concern for site management and policymakers (Shackley 2001).

In between indigenous and mainstream faiths, we find "cultural variants" of both where these traditions have met and melded (O'Brien and Palmer 2007). These "folk" religions represent a rich diversity of spiritual and cultural practices associated with sacred natural sites. Given that most of the sacred sites of mainstream faiths arose from natural features and associated cultural and spiritual practices, the fusion of local indigenous spirituality and mainstream religions contributes to ontological fluidity at the same time that it changes existing notions of indigeneity.

Adding to the diversity of sacred natural sites are those local landscapes that are deemed sacred and imbued with spirits by New Age and modern pagan movements (Harvey and Hardman 1996; Rountree 2014). Taylor (2009) coins the term "dark green religion" to refer to these movements

(Taylor 2009) and foresees an important and growing role for spiritual movements both in global environmental politics and in nature conservation. Through its rich and complex interaction with sacred places, dark green religion adds to the diversity of sacred natural sites and to the spectrum of management issues in nature conservation (e.g., Mallarach et al. 2012; Shackley 2001).

Sacred sites have been subject to scientific inquiry for many decades. They have aroused the curiosity of scientists in different disciplines, such as religious studies, the humanities, the social sciences, and the natural sciences (the last perhaps more recently, as the biodiversity values of sacred natural sites became evident and a case was built for their inclusion in the conservation agenda). In archaeology, cross-cultural studies of sacred natural sites have made clear that these are dynamic and resilient spaces that are often attached to living cultural traditions. An innovative exploration, building on insights by Carmichael et al. (1994), examines archaeologists' changing conception of sites' "significance," especially in the context of site classification, site recording, and subsequent site management. According to Carmichael et al. (1994: xiii), archaeologists defined sacred sites as "sites of special significance to people who created them, and/or those who now 'own,' investigate or protect them."

The wording of the definition suggests that the sacred dimensions of these places can also be of significance to those who research or manage them in cases in which the sites' creators are no longer present. Furthermore, Carmichael et al. (1994) suggest that learning from living ritual practices can improve understanding of the role of sacred places when the practice of ceremony and ritual have disappeared. These places pose significant challenges to site managers in terms of interpretation and conservation, especially when these are embedded in natural sites where there may be less evidence of previous human influence on the landscape (see Figure 1.1).

These challenges are also indicative of the practical and legal challenges involved in the protection, conservation, and revitalization of sacred natural sites that do have current custodians and where ceremony and rituals are performed according to spiritual traditions. In such cases, their conservation and management is often required to go hand in hand with rights-based approaches that help secure cultural management and religious practices in the face of external pressures. Often, these approaches are based on (or aspire to be) hard-won rights that assert indigenous peoples' right to self-determination and free, prior, and informed consent (FPIC), and they represent a struggle for freedom, respect, and reconciliation between dominant nation–states and indigenous cultures.

Figure 1.1. Cloch-Chearcal Agus Cairn (Dromberg Stone Circle) in County Cork, Ireland (photo credit: Bas Verschuuren, 2006)
This is one of Europe's ancient sacred sites and is now managed by a government institution. Located in County Cork in Ireland, it is the most frequented sacred site by tourists of all of Ireland, including those who use the site for the revival of ceremonial and spiritual purposes. This heavy, and unique, type of tourism poses challenges to the site's management and integrity thousands of years after its original custodians passed away (see Blain 2004: 241 for a broader discussion on this and Shackley 2001: 57–59) for a management-related example based on Iona Island).

For example, in Guatemala Oxlajuj Ajpop, the council of Mayan spiritual leaders seeks to secure indigenous management of sacred sites across the country in order to safeguard them from religious imposition, infrastructural development, and unsustainable forestry, mining, or conservation interventions. They have defined sacred sites as "naturally [occurring] or constructed places where cosmic energies are at a confluence to enable communication with ancestors for learning and practicing the spirituality, philosophy, science, technologies and art of the indigenous peoples" (Oxlajuj Ajpop 2008a: 11). From this definition it becomes clear that the meaning of sacred sites has to be understood from within Mayan ontologies, or "cosmovisions," as it is more commonly expressed by Mayan scholars themselves (Ivic de Monterrosso and Bravo 2008; see Figure 1.2 and the extract, "A Law on Sacred Sites in Guatemala," below).

Sacred Natural Sites in the Conservation Movement

There is mounting pressure on governments and corporate actors to increasingly ensure that the rights of indigenous peoples are recognized, and this struggle is also reflected in the modern conservation movement (Stevens 2010). The governance arrangements and management practices related to sacred natural sites also challenge conservationists to become better sensitized to how the worldviews of indigenous custodians interact with science-based conservation practices as well as frameworks of national and international law. In fact, as indigenous peoples become more influential in the international environmental policy arena, they are reshaping the debates and the conceptual underpinnings of the dominant conservation paradigm. Also, when legal incentives are absent, there exist good examples of conservationists and local people working together that are based on respectful and constructive forms of collaboration and partnership. These positive experiences also drive the growing recognition of sacred natural sites in the modern conservation movement.

Sacred natural sites gained importance in circles of cultural heritage and natural resource management during the late 1990s. Recommendation V.13, coming from the World Parks Congress held in Durban, South Africa in 2003, directed attention to the recognition of cultural and spiritual values and included sacred sites in global protected areas and conservation communities (IUCN 2003). The Durban Accord further opened a space for the recognition of sacred natural sites through the adoption of a new paradigm under which protected area laws, policies, governance, and management can be integrated equitably with the interests of all affected people (Brosius 2004).

International bodies such as the United Nations Educational, Scientific and Cultural Organization (UNESCO), the International Union for Conservation of Nature (IUCN), and the Convention on Biological Diversity (CBD) further developed their interest in sacred natural sites, which led to a series of international conferences (Schaaf 1998; Lee and Schaaf 2003; Schaaf and Lee 2005). The earliest attempt of conservationists to consolidate a working definition for sacred natural sites may be that of Oviedo and Jeanrenaud (2007: 77):

"Sacred natural sites include natural areas recognized as sacred by indigenous and traditional peoples, as well as natural areas recognized by institutionalized religions or faiths as places for worship and remembrance."

Later, for the purpose of the IUCN–UNESCO *Sacred Natural Sites Guidelines for Protected Area Managers*, sacred natural sites were defined as "areas of land or water having special spiritual significance to peoples and communities" (Wild and McLeod 2008: 7).

The IUCN World Commission on Protected Areas Specialist Group on Cultural and Spiritual Values of Protected Areas (CSVPA), which guided the development of these guidelines, also supported a coalition of organizations in the development of IUCN resolutions and recommendations. As a result, IUCN issued a resolution that focuses on sacred natural sites inside protected areas (IUCN 2008) as well as a recommendation on the protection of traditional governance systems related to sacred natural sites in protected areas and the wider landscape (IUCN 2012). The IUCN guidelines for protected area categories now recognize that sacred natural sites exist in all categories (I–VI) and across all governance types (comanaged protected areas, private protected areas, indigenous and local community conserved areas) of protected areas (Dudley 2008).

More recently, sacred natural sites were mentioned in the World Conservation Monitoring Centre's *Protected Planet Report* (Bertzky et al. 2012) and the United Nations Environmental Program's *Global Environmental Outlook 5* (UNEP 2012). Within the global conservation movement, sacred natural sites have gained recognition as the oldest conserved areas in the world (Dudley, Higgins-Zogib, and Mansourian 2009) and as nodes in important socioecological conservation networks (Verschuuren et al. 2010).

It also became apparent that not only do networks of sacred natural sites cover the lands of indigenous peoples in the so-called Third World, but they also extend into European and other developed countries. They do this as dormant remnants of past indigenous cultures that evaded, or were revitalized by, mainstream religions or were simply discovered anew by those seeking to reintroduce spirituality in nature (Taylor 2010; Rountree 2014).

The conservation management of sacred natural sites in technologically developed nations was initially assessed over the course of a series of workshops and case studies that took place under the auspices of the Delos Initiative (Mallarach and Papayanis 2007; Papayanis and Mallarach 2009; Mallarach et al. 2012). This resulted in a rich body of work that allows for creating specific guidelines for management and policy regarding the conservation of sacred natural sites of mainstream religions. However, the studies also show that the complex nature of folk religions and the ongoing influence of mainstream faiths in preexisting sacred natural sites would benefit from more detailed research and analysis—especially where a balanced approach to conservation is required.

Opportunities for Including Sacred Natural Sites in Conservation Policy and Practice

Protected areas and other conservation designations such as Indigenous and Community Conserved Areas (ICCAs), UNESCO World Heritage Sites and Biosphere Reserves, Ramsar Sites, and Globally Important Agricultural Heritage Systems create an important space in the policy and practice of conserving, restoring, and protecting sacred sites. Formal recognition that sacred sites form an interconnected and interdependent network is, however, lacking. Other efforts exploring sacred natural sites bring this aspect to light (e.g., Jonas et al. 2012; Verschuuren et al. 2010; Gaia Foundation 2012). This work also suggests that international and national policies be designed that adequately recognize traditional law and cultural practices related to sacred natural sites and their custodians. This would effectively support their protection, conservation, and revitalization as an interconnected network with a distinct spiritual dimension that is rooted in peoples' worldviews.

Various international treaties could be used for framing support for the protection of individual as well as interconnected networks of sacred natural sites and their custodians. Although reviewing these various international treaties would be a valuable exercise, more research and familiarization with jurisprudence in international and national policy would be required. A nonexhaustive list of key international instruments that could support the protection of sacred natural sites includes the following:

1. The Universal Declaration of Human Rights (1948)
2. The International Labour Organization's Convention (No. 169) concerning Indigenous and Tribal Peoples in Independent Countries (1989)
3. The World Heritage Convention—Cultural Landscapes (1992)
4. UNESCO's Man and the Biosphere's (MAB) Seville Strategy for Biosphere Reserves (1995)
5. UNESCO's Universal Declaration on Cultural Diversity (2001)
6. UNESCO's Convention for the Safeguarding of the Intangible Cultural Heritage (2003)
7. UNESCO's Convention on the Protection and Promotion of Diversity of Cultural Expressions (2005)
8. United Nations Declaration on the Rights of Indigenous Peoples (2007)
9. The Ramsar Convention on Wise Use of Wetlands (1971), especially Resolutions VIII.19 and IX.21

10. The Declaration on the Rights of Pacha Mama (2010)
11. The Convention on Biological Diversity (1992), especially Articles 8(j) and 10(c), the *Akwé Kon Guidelines*, the *Tkarihwaié:ri Code of Ethical Conduct*, and the community protocols under the Nagoya Protocol

This corpus of international legal and policy provisions arguably provides significant political leverage for the recognition and protection of sacred natural sites at the international level, through which signatory states are encouraged or mandated to enact similar provisions at the national level. Whether states are mandated or simply encouraged to do so depends on whether or not the international instrument in question is legally binding. For example, while the CBD (and the provisions contained therein) is legally binding, the United Nations Declaration on the Rights of Indigenous Peoples (UNDRIP) is voluntary. Nonetheless, UNDRIP is being used by indigenous peoples and local communities to pressure the CBD to adopt FPIC and encourage states to implement the *Akwé Kon Guidelines* as well as the *Tkarihwaié:ri Code of Ethical Conduct*.

Policy Spaces under the CBD for the Recognition of Sacred Natural Sites

In this section we expand on specific elements of the CBD, as this convention is focused on biodiversity conservation (the key objective of many conservationists), sustainable use, and equitable benefit sharing. In 2004, the United Nations Convention on Biological Diversity developed the *Akwé Kon Voluntary Guidelines for the Conduct of Cultural, Environmental and Social Impact Assessment Regarding Developments Proposed to Take Place on, or Which Are Likely to Impact on, Sacred Sites and on Lands and Waters Traditionally Occupied or Used by Indigenous and Local Communities* (Secretariat of the Convention on Biological Diversity 2004a). These guidelines are an excellent example of how sacred natural sites could be taken into account in relation to development and conservation activities. Because of the voluntary status of the *Akwé Kon Guidelines*, so far the government of Finland is the only CBD signatory that has applied and reported on the application of these guidelines (Juntunen and Stolt 2013).

In addition, the Tkarihwaié:ri Code of Ethical Conduct to Ensure Respect for the Cultural and Intellectual Heritage of Indigenous and Local Communities Relevant to the Conservation and Sustainable Use of Biological Diversity (Secretariat of the Convention on Biological Diversity 2011) helps create a space for various stakeholders to exchange interests and become mutually affected by each other's worldviews in a respectful and constructive manner.

Several articles under the convention, such as Articles 8(j) and 10(c), have great potential to legally support the restoration, protection, and conservation of sacred natural sites by their traditional custodians. Article 8(j) states that contracting parties should:

> Subject to its national legislation, respect, preserve and maintain knowledge, innovations and practices of indigenous and local communities embodying traditional lifestyles relevant for the conservation and sustainable use of biological diversity and promote their wider application with the approval and involvement of the holders of such knowledge, innovations and practices and encourage the equitable sharing of the benefits arising from the utilization of such knowledge, innovations and practices. (Secretariat of the Convention on Biological Diversity 1992: 7)

As Article 8j aims for respect, preservation, and maintenance of knowledge, innovations, and practices of indigenous people and local communities, it remains bound to national legislation. The article further suggests that parties promote the wider application of this knowledge with the approval and involvement of the knowledge-holders, but does not offer specific mechanisms for accomplishing this.

Another potentially important vehicle for the conservation of sacred natural sites is article 10c of the convention, which states, "Each Contracting Party shall, as far as possible and as appropriate, protect and encourage customary use of biological resources in accordance with traditional cultural practices that are compatible with conservation or sustainable use requirements" (Secretariat of the Convention on Biological Diversity 2009: 8). Here Article 10c provides an opening to recognize the diversity of protected area governance approaches, such as ICCAs and sacred natural sites. It also supports the full and effective participation of indigenous and local communities through the recognition, promotion, use, and application of traditional knowledge and traditional resource management practices. In relation to sacred sites, this also includes spiritual and religious teachings that are often an essential part of indigenous ontologies.

The CBD holds a particular challenge in this respect, as it obliges states to carry out reviews of national legislation and policies and implement reforms that recognize indigenous legal systems related to systems of governance and administration for land and waters, including sacred natural sites and other cultural sites. Two other provisions under the CBD could be marked as important motivators to this process, namely Aichi Target 11 (which sets an ambitious goal for bringing more land and sea under conservation) and the Nagoya Protocol (which regulates fair and equitable benefit sharing and suggests the implementation of community protocols).

Perhaps the most promising incentive under the CBD to improve the recognition of sacred natural sites may be provided under the Strategic Plan 2011–2020, which presents five strategic goals and twenty corresponding targets for safeguarding biodiversity. These targets are commonly known as the Aichi Targets.[1] Out of all the Aichi Targets, Target 11 is of specific importance in relation to the recognition of sacred natural sites and their custodians:

> By 2020, at least 17 per cent of terrestrial and inland water, and 10 per cent of coastal and marine areas, especially areas of particular importance for biodiversity and ecosystem services, are conserved through effectively and equitably managed, ecologically representative and well-connected systems of protected areas and other effective area-based conservation measures, and integrated into the wider landscapes and seascapes. (Secretariat of the Convention on Biological Diversity 2004b: 9)

Besides state-protected areas, there are many areas managed by nongovernmental organizations (NGOs), private landowners, communities, and indigenous peoples as well as faith groups and other custodians. Within the CBD, there is a growing recognition of the contributions made by ICCAs, many of which include sacred natural sites (Kothari et al. 2012). ICCAs have been broadly defined by the IUCN as "natural and/or modified ecosystems, containing significant biodiversity values, ecological benefits and cultural values, voluntarily conserved by indigenous peoples and local communities, through customary laws or other effective means" (Borrini and Oviedo 2004).

ICCAs, like many indigenous peoples' sacred natural sites, are based on human–nature relationships deeply rooted in cultural worldviews. One of the aspects that might differentiate sacred natural sites within and outside ICCAs could be that the "profound human–nature relationship" in sacred natural sites is characterized by heightened spiritual significance (Wild and McLeod 2008). Arguably, most sacred natural sites (like ICCAs) do qualify for "other effective area based conservation measures" and as such should be afforded recognition and appropriate protection under the CBD. This is important because a blanket protection for ICCAs would not automatically include all sacred natural sites, such as those of the established or newer religious communities.

Sacred Natural Sites, Effective Means for Conservation?

Sacred natural sites and ICCAs may be considered part of what the CBD recognizes as "other effective area-based conservation measures" under Aichi Target 11, but a reliable reference for how much land and sea is concerned does not currently exist (Berkes 2009). Approximately 80 percent of the world's biodiversity (Secretariat of the Permanent Forum on Indigenous Issues 2009) and 95 percent of the world's cultural diversity (Sobrevila 2008) is found on lands belonging to indigenous peoples and local communities—many of which are recognized as sacred or contain sacred sites.

The areas inhabited by indigenous peoples are often areas prioritized by biodiversity conservation organizations such as (1) biodiversity hotspots that contain most of the world's biodiversity and cover 2.3 percent of the Earth's surface, (2) megadiverse wilderness areas that cover 44 percent of the planet (Mittermeier 2003), and (3) the 12.7 percent of the world's terrestrial area and the 1.6 percent of the global ocean area that are currently designated by governments as protected areas (Bertzky et al. 2012). These designations show considerable overlap with indigenous territories, which cover approximately 20 percent of the world's surface, of which 7 percent is officially recognized by nation–states, while another estimated 13 percent remains unrecognized (Posey 2000).

Although they may potentially overlap, nature reserves and lands owned by religious institutions could potentially be included here, as they cover an estimated 7 percent of the world's surface (O'Brien and Palmer 2007). Many of those lands also contain sacred natural sites and pilgrimage routes of mainstream religions that connect sacred natural sites and can potentially play a tremendously important role in transboundary conservation. A good example of this is the Kailash Sacred Landscape Conservation and Development Initiative, which embodies a conservation effort between the governments of China, Nepal, India, and Pakistan.

The initiative also intersects with the spiritual values of local and indigenous people, as well as several major religious groups. They all journey on various pilgrimage trails to this mountain, which that is venerated by more than a billion Hindus and Muslims as the center of the universe and the abode of deities (Bernbaum 2010). Some of the pilgrimage routes and sacred natural sites may be looked after and governed by local communities or mainstream faiths, while many of them may also come under the jurisdiction of the government or private landowners.

There appears to be a clear trend toward increased recognition of many of the sensitivities involved in the conservation of sacred natural sites by international conservation organizations and conventions such as the IUCN, the World Conservation Monitoring Center (WCMC), the CBD, and the World Heritage Convention. These organizations argue that cultural and social safeguards, such as FPIC, need to complement biodiversity goals and that a start could be made by asserting and building on existing and emerging rights under national and international law. A good example of such social safeguards that would benefit from being made obligatory rather than voluntary is the CBD's *Akwé: Kon Guidelines.*

These cultural and social safeguards are particularly relevant in the context of the CBD's Programme of Work on Protected Areas, Element 2 on governance, participation, equity, and benefit sharing. Among other things, Element 2 includes Goal 2.2, which deals with enhancing and securing the involvement of indigenous and local communities and relevant stakeholders (Secretariat of the Convention on Biological Diversity 2004b). Most of the measures that were suggested under those goals were required to be implemented by the parties by 2008.

National Uptake and Implementation

In practice, it takes a long time for countries to translate the provisions of international conventions into appropriate, equitable, and effective national policy and legislation. In many countries, national policies and legislation on these matters are not being implemented, or they have a very limited degree of implementation. It is not uncommon that such changes are driven by the outcomes of specific national or international court cases. This is precious time during which sacred natural sites and their biocultural diversity are being lost.

Especially when it comes to improving the recognition for sacred natural sites and their custodians, lobbying international policy venues such as UNESCO and the CBD has to proceed hand in hand with advocacy efforts at the national level, often resulting from efforts at local, regional, and subnational levels. The National Biodiversity Strategic Action Plans (NBSAPs) under the Convention of Biological Diversity offer a particular opportunity for the integration of sacred natural sites to ensure that this strategy is implemented across all those sectors where activities can have an impact (positive and negative) on biodiversity.

National governments generally require a comprehensive overview and understanding of how international policies intersect with their national legislation (see below). In addition, a clear plan for their implementation can be extremely helpful in convincing specific government departments that the proposal is indeed feasible and has national value in terms of strengthening identity, values, and social cohesion. Of course, a law proposal will have to be preceded by a detailed process of local dialogue with custodians, communities, and knowledge holders as well as regional dialogue and consultation with interest groups, companies, and government departments.

Figure 1.2. Shu Sagrib-Al, sacred mountain in northwestern Guatemala (photo credit: Bas Verschuuren 2012)
This is a sacred mountain in northwestern Guatemala that the Mayans described in the *Popol Vuh*—the Mayan Holy Book—as the place of the awakening of the sun. The communities surrounding Shu Sagrib-Al, together with those of Oxlajuj Ajpop, bought the summit of the sacred mountain in order to protect it from threats such as mining and irresponsible forestry operations. The sacred site is also seen as the essential spiritual starting point for protecting and restoring the mountain's forests, waters, and the villages that depend on those. Oxlajuj Ajpop aims to support the restoration of many more sacred natural sites identified in the *Popol Vuh* by using the expertise of local communities as guidance for restoring biocultural landscapes.

A Law on Sacred Sites in Guatemala

In Guatemala, the National Law for Peace Agreements, signed in 1996, acknowledges the rights of indigenous peoples to practice their cultures in a specific territory and thereby implicitly acknowledges sacred (natural) sites as part of that territory. Since 1998, following the formation of the National Committee on the Protection of Sacred Sites, a law proposal regarding sacred sites was developed by Oxlajuj Ajpop—an organization of indigenous spiritual leaders representing Maya, Xinca, and Garifuna (Oxlajuj Ajpop 2008a). Many community dialogues were held and consolidated through developing a law proposal, which was accompanied by a national-level debate in parliament in support of it. The proposal answers the need for legal reforms that respect a socially and hence legally pluralistic state, cognizant of Article 66 of the Guatemalan Constitution: "The state must recognize, respect and promote the ways of life, customs, traditions, forms of social organization, the use of indigenous traditional dress, languages and dialects." (Political Constitution of the Republic of Guatemala 1985: 17).

The aim of the law proposal is to achieve recognition of community management, access, use, conservation, and administration of sacred sites by indigenous peoples. Article 1 states that the aim of the law proposal is to guarantee the historical, cultural, and spiritual rights of indigenous peoples by means of recognition, respect, dignification (the act of dignifying), use, conservation, administration, and access to sacred sites, constructed or natural, located in the national territory of Guatemala (Oxlajuj Ajpop 2008a). To inform the law proposal, more than twenty indigenous communities reflected on the importance of sacred sites, recorded and documented their histories, assessed their current ecological and legal status, and conducted celebrations to resanctify their sacred sites. They gained awareness of their rights to participate in the administration of sacred sites based on their indigenous management and governance systems and participated in dialogues about the law at the national level with congress and politicians, thus influencing the process on the basis of their shared experiences and understanding of their rights.

Adoption and implementation of the law proposal would enable the Mayan worldview to complement the contemporary Western state-based system currently adopted by the Guatemalan government. The law would arguably set a precedent to incorporate aspects of indigenous peoples' identities that are central to sacred natural sites into other areas of Guatemalan law such as education, natural resource management, health, and justice (see Oxlajuj Ajpop 2008b).

Positive experiences continue to encourage the development of new ways of working with custodians and the conservation of sacred natural sites. As these new ways emerge, they are best viewed as part of localized processes that may have a potential for outscaling and upscaling. From the ground up, various experiences derived from different countries may lead to the development of methods and approaches, such as guidelines or even toolkits. Examples of this are the "IUCN UNESCO Best Practice Guidelines No. 16" (Wild and McLeod 2008) and several statements developed by custodians of sacred natural sites themselves, such as the "Barcelona Statement" (Verschuuren et al. 2010), the "Statement on Common African Customary Laws for the Protection of Sacred Natural Sites" (Verschuuren and Wild 2012), and the Pyhätunturi Statement (2013) "Recognizing and Safeguarding Sacred Sites of Indigenous Peoples in Northern and Arctic Regions."

Biocultural Community Protocols (BCPs) are also a good example of a means to capture and upscale local needs in such a way that these can be communicated with external actors effectively (Bavikatte and Jonas 2009; Shrumm and Jonas 2012). These community protocols, intended to defend the community from particularly undesirable development interventions, are to some extent similar to the statements that have been developed by custodians of sacred natural sites; however, they expand on these by embedding specific issues of concern in a framework of traditional, national, and international law.

In learning from some of those localized experiences with BCPs that were created in order to protect sacred natural sites, the facilitators of these protocols conclude that they are best used in conjunction with a range of other tools and strategies to secure communities' rights, territories, and resources (see the extract, "Sacred Groves versus Gold Mining in Tancharra Community, Ghana," below). According to Booker and Shrumm (2012: 39), issues to consider when developing a BCP in relation to the protection and preservation of sacred natural sites include the following:

1. It is important to work effectively and appropriately with traditional authorities and custodians and in accordance with customary laws and values.
2. Careful contemplation of existing power dynamics is necessary in order to mitigate excessive influence of certain parties (including external parties).
3. It is critical to be mindful of competing views within communities and the effects of rights-based advocacy with respect to conflict with external actors (particularly in politically sensitive or repressive countries).
4. BCPs can be a lengthy process and are not a "quick fix"; rushing the BCP process can cause conflict and mistrust within communities, and care must be taken if communities face urgent or immediate concerns on their sacred sites.
5. Care must be taken when documenting sensitive community information: documentation can increase interest in natural resources or traditional knowledge by external parties.
6. BCPs can be used by external actors in unintended ways, such as coercing communities into agreements.

BCPs can be examples of what the CBD recognizes as community protocols under the *Nagoya Protocol on Access and Benefit Sharing of Biodiversity.* The Nagoya Protocol is a supplementary agreement to the CBD and provides a transparent legal framework for the fair and equitable sharing of benefits arising out of the utilization of genetic resources. This international recognition of community protocols has in turn boosted the local uptake of the tool as a means to assist local and indigenous peoples with communicating and with advocating their interests to external actors—something that is also very useful in the context of many sacred natural sites. Although community protocols in themselves are not new, many have been developed by communities over time for various different reasons, and the Nagoya Protocol effectively supports their wider development and application.

The "Bio Cultural Community Protocol Toolkit" (Shrumm and Jonas 2012) has been developed based on learning from many different local experiences—many involving sacred natural sites—gathered from around the world. The example below shows how capacity building can be achieved by building on endogenous values of the communities themselves. In doing so, the Tancharra community and a Ghanaian NGO—the Center for Indigenous Knowledge and Organizational Development (CIKOD)—have worked together to protect its sacred groves from destruction by an Australian gold-mining company.

Sacred Groves versus Gold Mining in Tancharra Community, Ghana (source: Guri et al. 2012)

The Tancharra community—in the Upper West Region of Ghana—and a Ghanaian NGO, the Center for Indigenous Knowledge and Organizational Development (CIKOD), work together to protect the communities' sacred groves from destruction by an Australian gold-mining company that has been given a concession by the Ghanaian government to mine in the area without having to consult with the local communities that live there.

Traditionally, regulations for the protection of the sacred groves are enforced by the Tingandem, who are spiritual leaders as well as keepers of the land and adviser to the chief and the *pognaa,* the chief's female counterpart. The Tingandem formulated a statement protesting the activities of the miners and asked the government to safeguard their sacred groves and sites from both legal and illegal mining. All the Tingandem appended their thumbprints to this paper and requested CIKOD to send this to the appropriate authorities for their attention and action.

The Tancharra community has become a pilot study in the African Biocultural Community Protocol Programme but CIKOD had already been working with the Tancharra community, helping them to strengthen their self-reliance and ability to sustain their development "from within." To this end, CIKOD has developed and employed a series of tools known as Community Organizational Development Tools, which include the following:

—Community Institutions and Resources Mapping
—Community Visioning and Action Planning
—Community Organizational Self-Assessment
—Community Institutional Strengthening
—Learning-Sharing and Assessment

The Tancharra community members were trained to use these tools with the aim of strengthening their capacity to respond to issues of importance, in this case the gold-mining activities. The biocultural community protocol builds on these capacities by including a statement on the communities' traditional knowledge and practices related to their resources in order to protect them within the framework of customary, national, and international legislation. The BCP also sets out community controls for how to negotiate any commercial agreement with outsiders and calls on the mining company, the government, and other stakeholders to save the community's sacred groves from the impacts of gold prospecting and mining.

The preparation of the protocol has been a tremendously empowering activity for the Tancharra community. Strengthened in their means for self-determination, the traditional leaders, especially the Tingandem, played a key role in developing the protocol and in discussions within the community. This process also helped the community become aware of their legal rights, such as the right to free and prior informed consent enshrined in the United Nations Declaration on the Rights of Indigenous Peoples. Initially the community used the protocol to negotiate a moratorium on mining for a year, but since the operations affect the sacred groves of all the communities within the mining concession CIKOD is expanding the work on BCPs to other communities in the region and engages them in regional-level debate and advocacy.

Toward a Middle Ground

Keeping in mind that sacred natural sites are part of peoples' worldviews, there is the risk that a dissection of biodiversity-related knowledge would isolate it from its spiritual meaning, which would arguably be indivisible in traditional lifestyles. The problems of isolating traditional ecological knowledge from its context (or the worldview on which it rests) for business or conservation management purposes has been well documented (Berkes 1999; Berkes and Turner 2006). Among scholars in political ontology, this separation or isolation is perceived as one of the outcomes of ontological power imbalances, or those imbalances between the ontologies of indigenous peoples and Western planners and conservationists (Hunt 2014).

Moving beyond traditional ecological knowledge, the custodians of sacred natural sites often mediate between the human and the spirit world. Hence, sacred natural sites can be seen as places for mediation and guidance and, in some cases, as sources of traditional law. Because sacred natural sites depend on institutions that have the de facto and/or the de jure capacity to develop and enforce decisions based on traditional or cultural laws, they also need to be seen as part of the ontological fabric of indigenous peoples.

Because of the different assumptions about "the Other," it is not uncommon that indigenous peoples and conservationists (many of whom have been trained in Western institutions) have different ideas about what concepts such as "conservation" and "sustainable development" might mean. Cultural anthropologists (Blaser 2009; West 2006) have repeatedly shown that when both parties evaluate the outcomes of the conservation projects in which they participated, they both feel that the other party falls short in delivering conservation and development. In such cases, the different understandings and conceptualizations of conservation and development require sensitizing and understanding on both sides in order for both parties to meaningfully work together.

This does not mean that an agreement needs to be reached on what constitutes the "true" meaning of conservation and development. Misinterpretation and distortion may be difficult if not impossible to rule out; yet, as White (1991: x) observes from a historical perspective, "From these misunderstandings arise new meanings and through them new practices—the shared meanings and practices of the middle ground."

In this respect, the very exercise of policy making is influenced mostly by dominant ontologies, while others have traditionally been excluded. This is of particular importance for indigenous peoples and sacred natural sites custodians, who would benefit from the creation and identification of spaces in international policies where they can bring their worldviews to the fore.

Within both the management and policy context, the recognition of sacred natural sites can also have undetermined and potentially negative side effects. One of them would be the inclusion of sacred sites in management or policy measures to which their custodians would not normally consent. Many sacred natural sites are culturally sensitive areas controlled by decentralized local or religious governance systems that may not have the capacity, knowledge, or means to facilitate appropriate linkages to existing national or international policy actions.

Woodley et al. (2012: 24) also recognize these sensitivities in relation to the recent creation of the ICCA Registry[2] maintained by the World Conservation Monitoring Institute (WCMC): "Custodians of some ICCAs and sacred natural sites may have good reasons for not wanting to appear on an international database, because it could draw increased attention to sites that retain value in part because of their isolation." Carmichael et al. (1994)

also mention the "secret sacred" and accordingly the need to control and manage knowledge related to sacred sites.

It is important to respect cultural protocols and realize that specific knowledge related to sacred natural sites is also regulated through cultural and social rules and regulations that are not readily accessible to outsiders such as companies, conservationists, and researchers. This is of particular importance to the World Database on Sacred Natural Sites (SANASI),[3] which, like the ICCA registry, makes use of a FPIC procedure to check whether information submitted is culturally harmful for custodians and communities: it should be up to them to decide what information can be shared and what needs to remain within the community (Corrigan and Hay-Edie 2013).

Appropriate recognition and innovative biocultural conservation approaches are required for the protection and conservation of sacred natural sites, especially if these are to help achieve the CBD's goal of increasing the area of land under protected areas and other effective means to 17 percent by 2020 (Bertzky et al. 2012). Networks of resilient, adaptive, and effectively managed sacred natural sites are thought to be able to significantly contribute to this global conservation mission. In many cases, the reality is that sacred natural sites require improved protection, conservation, and restoration efforts. In order to recognize and restore this network of sacred natural sites to its potential conservation value, a measure of its contribution to the global conservation target can work as an incentive. Given the cultural sensitivities, social safeguards, such as FPIC and full compliance with the principle of self-determination, will need to be observed in order to avoid adverse impacts on these sensitive places and their custodians.

The protection, conservation, and revitalization of sacred natural sites can be enabled within the framework of local, national, and international laws. However, the praxis, management, and policy engagement at each of these levels should be scaled to purpose, bearing in mind the need for local custodians to exercise their cultural, spiritual, and religious responsibilities. Conservation organizations can also help to raise the bar with local, regional, and national governments by demonstrating "best practices" in this regard. To date, however, a review of existing laws and policies (international and national) that assist with the conservation of sacred natural sites and faith-based conservation areas is still lacking. Such an effort would include a review of existing rights as well as regulations and policies that intersect with laws that help protect sacred natural sites; it would also need to include community conserved areas and faith-based conservation areas.

A promising initiative in this regard is the existing review focusing on ICCAs by Jonas (2012), whose work offers a synthesis based on the strengths and weaknesses of legislation in eighteen countries in the Pacific; South and Central Asia; East, West, and South Africa; and Europe and the Americas. The review makes a distinct effort to look at sacred natural sites in each of these regions. As sacred natural sites are complex and often intersect with various pieces of legislation, not all legal representatives performing the review were able to focus on them throughout the research and instead prioritized the use of their limited resources to focus on the concept of territory. Additional and more comprehensive analysis covering a full suite of rights—including cultural and religious—will be required for sacred natural sites, landscapes, and pilgrimage routes.

Biocultural conservation approaches are required to effectively combine the collective practices, knowledge, and wisdom of the custodians of sacred natural sites with contemporary conservation management and policy. Several initiatives are working to advance the conservation of sacred natural sites. Their activities have focused on supporting individual groups of custodians or specific types of sacred natural sites in selected regions in the world. Nonetheless, they developed a set of strategic directions (each followed by a number of subactions) to guide the most necessary actions for the conservation of sacred natural sites (Verschuuren et al. 2010):

1. Support the autonomous protection and management of sacred natural sites by their custodians.
2. Reduce the threats and halt the loss of sacred natural sites.
3. Support cultural revitalization and the strengthening of communities and their connections with their sacred natural sites.
4. Increase understanding and awareness, particularly at the national level, of the importance and role of sacred natural sites and promote the formation of appropriate national policies and laws.
5. Build up a body of increased knowledge of sacred natural sites, using different ways of knowing, including traditional knowledge, holistic science, the arts, and media.
6. Access and generate funding for sacred natural sites identifying a diversity of resources (financial and otherwise) to support sacred natural sites.

In order to effectively promote the recognition for the protection, conservation, and revitalization of sacred natural sites, collaborative efforts to implement these strategic actions are required by custodians, communities, scientists, conservationists, and civil society organizations.[4]

Figure 1.3. Mphathe Makaulule from Venda, South Africa, speaking at the IUCN World Conservation Congress in South Korea, 2012 (photo credit: Bas Verschuuren 2012)
Mphathe Makaulule speaks about her experiences, stressing the importance of internationally recognized guidelines, which can help local custodians achieve recognition for their cause of protecting, conserving, and revitalizing sacred natural sites.

Sacred natural sites are increasingly recognized for their contribution to the conservation and sustainable use of both global biodiversity and tangible and intangible cultural diversity. Initially, a small group of enthusiastic conservation professionals signaled the need to give more importance to sacred natural sites in the realm of conservation. They started working together with indigenous peoples, policy makers, and relevant institutions to achieve a better understanding of the issues that are critical to their conservation. This work developed into a set of guidelines for protected area managers (Wild and McLeod 2008), which can help construct a middle ground for sacred natural site custodians and conservationists. The guidelines include the following practical steps for management and planning (Wild and McLeod 2008: 21):

Principle 1: Recognize sacred natural sites already located in protected areas.

Principle 2: Integrate sacred natural sites located in protected areas into planning processes and management programs.
Principle 3: Promote stakeholder consent, participation, inclusion, and collaboration.
Principle 4: Encourage improved knowledge and understanding of sacred natural sites.
Principle 5: Protect sacred natural sites while providing appropriate management access and use.
Principle 6: Respect the rights of sacred natural site custodians within an appropriate framework of national policy.

The guidelines are written in a manner that is cognizant of the importance of traditional custodians in managing their sites. They clearly recognize the duties and rights of custodians and announce that it would be inappropriate for organizations intervening from outside to provide management advice regarding sacred sites without the permission of, and advice from, the appropriate custodians who have often successfully cared for these sites for many generations. The guidelines have been developed to promote cooperation between protected area managers and custodians with the goal of enhanced protection and conservation of these special places.

Since their launch, the guidelines have been published in seven languages: English, Russian, Spanish, Estonian, French, Korean, and Japanese. The essential parts of the guidelines have also been translated into Italian and Persian; increasingly, park services, local people, and dedicated individuals express an interest in translating the guidelines into their respective languages with the desire to see them implemented. A good example of the national uptake of the guidelines is the Estonian translation, which was undertaken as part of a broader national strategy of sensibilization toward (or heightened awareness of) and inventory of sacred natural sites (Kaasik 2011).

Because of the recognition of the duties and rights of traditional custodians, traditional custodians seeking to engage with environmental or protected area authorities and other external actors can also make use of the guidelines. For example, the Yolngu people in northeast Arnhem Land, Australia, and the Dhimurru Aboriginal Corporation have incorporated the guidelines into their Cultural Heritage Management Plan as part of their Indigenous Protected Area. Others, such as the Venda people in South Africa, used the guidelines as an international standard to convince their government that their sacred sites were to be taken seriously. The custodians of the Venda clans' sacred sites, the Makhadzhi, formed a committee called Dzomo la Mupo (Voices of Earth), which filed and won a lawsuit against a party that had initiated the development of a tourism resort on the grounds surrounding their sacred waterfall (see Figure 1.3). The Venda attracted the government's attention and are now pushing to have their sacred natural sites recognized as an interdependent network.

The guidelines also call on protected area managers and conservationists themselves to advocate appropriate, relevant policy changes that will improve the management of sacred natural sites locally, nationally, and globally. The guidelines generally develop from the specific and the local to the more general and national level. In this manner they retain a high level of endogenous experience at the core of their implementation. To ensure that these locally relevant specificities do not get lost in overarching management decisions taken at the regional or national scale, it is recommended that individual protected area managers advocate appropriate, relevant policy changes that will help improve the management of sacred natural sites from the local to the national and even international level. One way of enabling this type of experience with the guidelines is through a national strategy that challenges protected area managers, heritage professionals, and policymakers alike to engage in the construction of a middle ground that overcomes the pitfalls of the past.

Bas Verschuuren, M.Sc., is a freelance adviser and researcher working on biocultural conservation and rural development issues. He is an associate researcher at Wageningen University, cochair to the International Union for Conservation of Nature–World Commission on Protected Areas Specialist Group on Cultural and Spiritual Values of Protected Areas, and cofounder of the Sacred Natural Sites Initiative, www.sacrednaturalsites.org. He is also lead editor of *Sacred Natural Sites: Conserving Nature and Culture.*

Robert Wild, M.D., is a coordinator for the Sacred Natural Sites Initiative, a freelance conservation consultant, and a biodiversity adviser, Edinburgh, U.K.

Gerard Verschoor, Ph.D., is an assistant professor in the Sociology of Development and Change Group, Wageningen University, the Netherlands. He is an expert on the nexus of conservation and development, specifically the communicative and/or conceptual disjunctures that arise when different "worlds" or ontologies (most notably those of conservationists and indigenous people) meet.

Notes

1. For a comprehensive overview of the Aichi Targets, see www.cbd.int/sp/targets.
2. See www.iccaregistry.org.
3. Available at www.sanasi.org.
4. Sea www.sacrednaturalsites.org.

References

Bavikatte, Kabir and Harry Jonas, eds. 2009. *Bio-cultural Community Protocols: A Community Approach to Ensuring the Integrity of Environmental Law and Policy.* Nairobi: UNEP.

Berkes, Fikret. 1999. *Sacred Ecology: Traditional Ecological Knowledge and Resource Management.* New York: Routledge.

Berkes, Fikret. 2009. "Community Conserved Areas: Policy Issues in Historic and Contemporary Context." *Conservation Letters* 2, no. 1: 20–25.

Berkes, Fikret and Nancy Turner. 2006. "Knowledge, Learning and the Evolution of Conservation Practice for Social-Ecological System Resilience." *Human Ecology* 34, no. 4: 479–494.

Bernbaum, Edwin. 2010. "Sacred Mountains and Global Changes: Impacts and Responses." In *Sacred Natural Sites: Conserving Nature and Culture,* edited by Bas Verschuuren, Robert Wild, Jeffrey McNeely, and Gonzalo Oviedo, 33–41. London: Earth Scan.

Bertzky, Bastian, Colleen Corrigan, James Kemsey, Siobhan Kenney, Corinna Ravilious, Charles Besançon, and Neil Burgess. 2012. *Protected Planet Report 2012: Tracking Progress towards Global Targets for Protected Areas.* Gland, Switzerland: IUCN; Cambridge: UNEP-WCMC.

Blain, Jenny and Robert J. Wallis. 2004. "Sacred Sites, Contested Rites/Rights Contemporary Pagan Engagements with the Past." *Journal of Material Culture* 3: 237–261.

Blaser, Mario. 2009. *American Anthropologist* 11, no. 1: 10–20.

Booker, Stephanie and Holly Shrumm. 2012. "Protecting the Sacred: The Role Community Protocols Play in the Protection of Sacred Natural Sites." In *Sacred Natural Sites: Sources of Biocultural Diversity,* edited by Bas Verschuuren and Robert Wild, 80. Salt Spring City: Terralingua.

Borrini, Grazia and Gonzalo Oviedo. 2004. *Indigenous and Local Communities and Protected Areas: Towards Equity and Enhanced Conservation: Guidance on Policy and Practice for Co-managed Protected Areas and Community Conserved Areas.* Gland, Switzerland: IUCN.

Brosius, J. Peter. 2004. "Indigenous Peoples and Protected Areas at the World Parks Congress." *Conservation Biology* 18, no. 3: 609–612.

Büscher, Bram, Sian Sullivan, Katja Neves, Jim Igoe, and Dan Brockington. 2012. "Towards a Synthesized Critique of Neoliberal Biodiversity Conservation." *Capitalism, Nature, Socialism* 23, no. 2: 4–30.

Byrne, Dennis. 2010. "The Enchanted Earth; Numinous Sacred Sites." In *Sacred Natural Sites, Conserving Culture and Nature,* edited by Bas Verschuuren, Robert Wild, Jeffrey McNeely, and Gonzalo Oviedo., pp. 53–61. London: EarthScan.

Carmichael, David L., Jane Hubert, Brian Reeves, and Audhild Schanche. 1994. *Sacred Sites, Sacred Places.* New York and Oxon: Routledge.

Chape, Stuart, Mark Spalding, and Martin Jenkins. 2008. *The World's Protected Areas, Status, Values and Prospects in the 21st Century.* Berkeley: University of California Press.

Corrigan, Colleen and Terence Hay-Edie. 2013. *A Toolkit to Support Conservation by Indigenous Peoples and Local Communities: Building Capacity and Sharing Knowledge for Indigenous Peoples' and Community Conserved Territories and Areas (ICCAs).* Cambridge, UK: UNEP-WCMC.

Delgado, Freddy, Cesar Escobar, Bas Verschuuren, and Wim Hiemstra. 2010. "Sacred Natural Sites, Biodiversity and Well-Being: The Role of Sacred Sites in Endogenous Development in the COMPAS Network." In *Sacred Natural Sites, Conserving Culture and Nature,* edited by Bas Verschuuren, Robert Wild, Jeffrey McNeely, and Gonzalo Oviedo, pp. 188–197. London: EarthScan.

Dudley, Nigel, ed. 2008. *Guidelines for Applying Protected Area Management Categories.* Gland, Switzerland: IUCN.

Dudley, Nigel, Liza Higgins-Zogib, and Stefanie Mansourian. 2009. "Links between Protected Areas, Faiths, and Sacred Natural Sites." *Conservation Biology* 23, no. 3: 568–577.

Gaia Foundation. 2012. *Legal Recognition of Indigenous Sacred Natural Sites and Territories: Community-Led Strategies for Gaining Legal Recognition for Indigenous Sacred Natural Sites and Territories and Their Related Custodial Governance Systems.* Draft version prepared for the IUCN World Conservation Congress, 5–14 October 2008. Barcelona. London and Bogota, Gaia Foundation.

Gomez Felipe, Wim Hiemstra, and Bas Verschuuren. 2010. "A Law Proposal on Sacred Sites in Guatemala." *Policy Matters* 17: 119–123.

Guri, Bernard, Daniel Banuoko, Emanuel Derbile, Wim Hiemstra, and Bas Verschuuren. 2012. "Ghanaian Communities Organize Their Own Development and Protect Sacred Groves from Gold Mining Using Biocultural Community Protocols." *PLA* 65: 21–130.

Harvey, Graham and Charlotte Hardman, eds. 1996. *Paganism Today: Wiccans, Druids, the Goddess and Ancient Earth Traditions for the Twenty-First Century.* New York: Thorsons/HarperCollins.

Hunt, Sarah. 2014. "Ontologies of Indigeneity: The Politics of Embodying a Concept." *Cultural Geographies* 21, no. 1: 27–32.

International Labour Organisation. 1991. *Convention No. 169 Concerning Indigenous and Tribal Peoples in Independent Countries.* San José: ILO Office for Central America, Panama, and the Dominican Republic.

International Society of Ethnobiology. 2006. "International Society of Ethnobiology Code of Ethics (with 2008 Additions)." Available at: http://ethnobiology.net/code-of-ethics/ (accessed April 2014).

IUCN. 2003. *Recommendations: Vth IUCN World Parks Congress, Durban South Africa 2003.* Gland, Switzerland: IUCN. (accessed April 2014).

IUCN. 2008. "Resolution 4.038: Recognition and Conservation of Sacred Natural Sites in Areas." Available at: http://sacrednaturalsites.org/wp-Protected content/uploads/2011/09/IUCN-Resolution-4038-Recognition-and-conservation-of-sacred-natural-sites-in-protected-areas.pdf (accessed April 2014).

IUCN. 2012. "Recommendation M054: Sacred Natural Sites—Support for Custodian Protocols and Customary Laws in the Face of Global Threats and Challenges." Available at: http://sacrednaturalsites.org/wp-content/uploads/2013/01/Resolution-2012-054-Sacred-Natural-Sites-EN.pdf (accessed April 2014).

Ivic de Monterroso, Matilde and Iván Azurdia Bravo, eds. 2008. *Ciencia y Tecnica Maya: Con la Colaboracion del Consejo Nacional de Ancianos Principles y Guias Espirituales Mayas, Xincas y Garifunas.* Guatemala City: Fundación Solar.

Jonas, Harry, Ashish Kothari, and Holly Shrumm. 2012. *Legal and Institutional Aspects of Recognising and Supporting Conservation by Indigenous Peoples and Local Communities: An Analysis of International Law, National Legislation, Judgements and Institutions as They Interrelate with Territories and Areas Conserved by Indigenous Peoples and Local Communities.* Bangalore, Pune, and Delhi: Natural Justice and Kalpavriksh.

Juntunen, Suvi and Elina Stolt, eds. 2013. *Application of Akwé: Kon Guidelines in the Management and Land Use Plan for the Hammastunturi Wilderness Area.* Vantaa: Metsähallitus Natural Heritage Services.

Kaasik, Atho. 2011. "Conserving Sacred Natural Sites in Estonia." In *The Diversity of Sacred Lands in Europe: Proceedings of the Third Workshop of the Delos Initiative, Inari/Aanaar, Finland, 1–3 July 2010,* edited by Josep-Maria Mallarach, Thymio Papayannis, and Rauno Väisänen, 292. Gland, Switzerland and Vantaa, Finland: IUCN and Metsähallitus Natural Heritage Services.

Kothari, Ashish, Colleen Corrigan, Harry Jonas, Aurélie Neumann, and Holly Shrumm, eds. 2012. *Recognizing and Supporting Territories and Areas Conserved by Indigenous Peoples and Local Communities: Global Overview and National Case Studies.* Technical Series no. 64. Montreal: Secretariat of the Convention on Biological Diversity.

Lee, Cathy and Thomas Schaaf, eds. 2003. "The Importance of Sacred Natural Sites for Biodiversity Conservation." In *Proceedings of the International Workshop on the Importance of Sacred Natural Sites for Biodiversity Conservation; Kunming and Xishuangbanna Biosphere Reserve, People's Republic of China,* 17–20. Paris: UNESCO.

Mallarach, Josep-Maria, ed. 2012. *Spiritual Values of Protected Areas of Europe: Workshop Proceedings; Workshop Held 2–6 November 2011 at the International Academy for Nature Conservation on the Isle of Vilm, Germany.* Bonn: BfN Bundesamt für Naturschutz.

Mallarach, Josep-Maria and Thymio Papayannis, eds. 2007. *Protected Areas and Spirituality. Proceedings of the First Workshop of the Delos Initiative, Monastery of Montserrat, Catalonia, Spain, 24–26 November 2006.* Barcelona: Publicaciones de l'Abadia de Montserra; Gland, Switzerland: World Conservation Union; Barcelona: Fundació Territori i Paisatge; Catalonia: Government of Catalonia Department of the Environment and Housing.

Mallarach, Josep-Maria, Thymio Papayannis, and Rauno Väisänen, eds. 2012. *The Diversity of Sacred Lands in Europe. Proceedings of the Third Workshop of the Delos Initiative, Inari/Aanaar Finland, 1–3 July 2010.* Gland, Switzerland and Vantaa, Finland: IUCN and Metsähallitus Natural Heritage Services.

Mathez-Stiefel, Sarah-Lan, Sébastien Boillat, and Stephan Rist. 2007. "Promoting the Diversity of Worldviews: An Ontological Approach to Biocultural Diversity." In *Endogenous Development and Bio-cultural Diversity: The Interplay of Worldviews, Globalisation and Locality,* edited by Bertus Haverkort and Stephan Rist, 67–82. Compas Series on Worldviews and Science, No. 6. Leusden, Netherlands: Compas.

Mittermeier, Russell A., Cristina G. Mittermeier, Thomas M. Brooks, John D. Pilgrim, William R. Konstant, Gustavo A. B. Fonseca, and Cyril Kormos. 2003. "Wilderness and Biodiversity Conservation." *PNAS 100, no.* 18: 10309–10313.

O'Brien, Joanne and Martin Palmer. 2007. *The Atlas of Religion: Mapping Contemporary Challenges and Beliefs.* London: EarthScan.

Oviedo, Gonzalo and Sally Jeanrenaud. 2007. "Protecting Sacred Natural Sites of Indigenous and Traditional Peoples." In *Protected Areas and Spirituality. Proceedings of the First Workshop of the Delos Initiative, Montserrat, 23–26 November 2006,* edited by Josep-Maria Mallarach and Thymio Papayannis, 77–99. Gland, Switzerland: IUCN; Montserrat: Publicaciones de l'Abadia de Montserrat.

Oxlajuj Ajpop. 2008a. *Initiative in Support of the Law on Sacred Natural Sites of Indigenous Peoples No. 3835, On the Occasion of the New Mayan Cycle Oxlajuj B'aqtun.* Guatemala City: Conferencia Nacional Oxlajuj Ajpop.

Oxlajuj Ajpop. 2008b. *Organización Para La Recuperación, Conservación, Dignificación y Administración de los Lugares Sagrados Mayas.* Guatemala City: Conferencia Nacional Oxlajuj Ajpop.

Oxlajuj Ajpop. 2009. *Agenda Socio Ambiental desde el Pensamiento de los Pueblos Indígenas por los Derechos de la Madre Tierra.* Guatemala City: Compas Meso America.

Papayannis, Thymio and Josep-Maria Mallarach, eds. 2009. *The Sacred Dimension of Protected Areas. Proceedings of the Second Workshop of the Delos Initiative—Ouranoupolis 2007.* Gland, Switzerland: IUCN; Athens: Med-INA.

Political Constitution of the Republic of Guatemala. 1985. *Political Constitution of the Republic of Guatemala. As Amended by Legislative Decree No. 18–93 of 17 November 1993.* Guatemala: National Constituent Assembly of the Republic of Guatemala.

Posey, Darrell Addison. 1999. *Cultural and Spiritual Values of Biodiversity: A Comprehensive Contribution to the UNEP Global Biodiversity Assessment.* London: Intermediate Technology Publications/UNEP.

Posey, Darrell Addison. 2000. "The 'Balance Sheet' and the 'Sacred Balance': Valuing the Knowledge of Indigenous and Traditional Peoples." *World Views: Global Religions, Culture, and Ecology* 2, no. 2: 91–106.

Pyhätunturi Statement. 2013. "Recognizing and Safeguarding Sacred Sites of Indigenous Peoples in Northern and Arctic Regions." Conference Statement and

Recommendations from the International Conference "Protecting the Sacred: Recognition of Sacred Sites of Indigenous Peoples for Sustaining Nature and Culture in Northern and Arctic Regions" held in Pyhätunturi and Rovaniemi, Finland, 11–13 September 2013.

Rountree, Kathryn. 2014. "Neo Paganism, Native Faith and Indigenous Religion: A Case Study of Malta within the European Context." *Social Anthropology* 22, no. 1: 81–100.

Schaaf, Thomas. 1998. *Report on the Workshop on Natural Sacred Sites: Cultural Integrity and Biological Diversity* (unpublished report). Workshop held at UNESCO Paris, 22–25 September 1998. Paris: UNESCO.

Schaaf, Thomas and Cathy Lee. 2006. "The Role of Sacred Natural Sites and Cultural Landscapes." In *UNESCO-MAB Proceedings of the Tokyo Symposium,* 30 May–2 June 2005, Tokyo, 341. Paris: United Nations Educational, Scientific and Cultural Organization.

Secretariat of the Convention on Biological Diversity. 2004a. *Akwé: Kon Voluntary Guidelines for the Conduct of Cultural, Environmental and Social Impact Assessment Regarding Developments Proposed to Take Place on, or Which Are Likely to Impact on, Sacred Sites and on Lands and Waters Traditionally Occupied or Used by Indigenous and Local Communities.* CBD Guidelines Series. Montreal: Secretariat of the Convention on Biological Diversity.

Secretariat of the Convention on Biological Diversity. 2004b. "Programme of Work on Protected Areas (CBD Programmes of Work)." Montreal: Secretariat of the Convention on Biological Diversity. Available at: https://www.cbd.int/doc/publications/pa-text-en.pdf (accessed 26 February 2014).

Secretariat of the Convention on Biological Diversity. 1992. "Convention on Biological Diversity." Available at: http://www.cbd.int/doc/legal/cbd-en.pdf (accessed 26 February 2014).

Secretariat of the Permanent Forum on Indigenous Issues. 2009. *State of the World's Indigenous Peoples Department of Economic and Social Affairs Division for Social Policy and Development.* New York: United Nations.

Secretariat of the Convention on Biological Diversity. 2011. "Tkarihwaié:ri Code of Ethical Conduct to Ensure Respect for the Cultural and Intellectual Heritage of Indigenous and Local Communities Relevant to the Conservation and Sustainable Use of Biological Diversity." Available at: http://www.cbd.int/traditional/code.shtml (accessed February 2014).

Shackley, Myra. 2001. *Managing Sacred Sites: Service Provision and Visitor Experience.* Cengage Learning EMEA. London: Continuum.

Sheridan, Michael J. and Celia Nyamweru, eds. 2008. *African Sacred Groves: Ecological Dynamics and Social Change.* Oxford: James Currey.

Shrumm, Holly and Harry Jonas, eds. 2012. *Biocultural Community Protocols: A Toolkit for Community Facilitators.* Cape Town: Natural Justice.

Sobrevila, Claudia. 2008. *The Role of Indigenous Peoples in Biodiversity Conservation: The Natural but Often Forgotten Partners.* Washington, D.C.: World Bank.

Stevens, Stan. 2010. "Implementing the UN Declaration on the Rights of Indigenous Peoples and International Human Rights Law through the Recognition of ICCAs." *Policy Matters* 17: 181–194.

Taylor, Bron Raymond. 2009. *Dark Green Religion: Nature Spirituality and the Planetary Future.* Berkeley: University of California Press.

Theodoratus, Dorothea J. and Frank LaPena. 1994. "Wintu Sacred Geography of Northern California." In *Sacred Sites, Sacred Places,* edited by David Carmichael, Jane Hubert, Brian Reeves, and Audhild Schance, 20–31. London: Routledge.

UNDRIP. 2007. *United Nations Declaration on the Rights of Indigenous Peoples.* UN Document A/RES/61/295.

UNEP. 2012. *Global Environmental Outlook 5, Environment for the Future We Want.* Nairobi: United Nations Environmental Programme.

United Nations. 2011. *The UN Guiding Principles on Business and Human Rights: Implementing the United Nations; "Protect, Respect and Remedy" Framework.* New York and Geneva: United Nations, Human Rights, Office of the High Commissioner.

United Nations Educational, Scientific and Cultural Organization. 2003. "Convention for the Safeguarding of Intangible Cultural Heritage." Available at: http://portal.unesco.org (accessed 23 August 2010).

Verschuuren, Bas. 2010. "Arguments for Developing Biocultural Conservation Approaches for Sacred Natural Sites." In *Sacred Natural Sites, Conserving Culture and Nature,* edited by Bas Verschuuren, Robert Wild, Jeffrey McNeely, and Gonzalo Oviedo, 62–72. London: EarthScan.

Verschuuren, Bas and Robert Wild. 2012. "Safeguarding Sacred Natural Sites: Sustaining Nature and Culture." In *Sacred Natural Sites: Sources of Biocultural Diversity,* edited by Bas Verschuuren and Robert Wild, 6–7. Terralingua: Salt Spring Island and Sacred Natural Sites Initiative.

Verschuuren, Bas, Robert Wild, and Jeffrey McNeely. 2010. "Introduction: Sacred Natural Sites, the Foundations of Conservation." In *Sacred Natural Sites, Conserving Culture and Nature,* edited by Bas Verschuuren, Robert Wild, Jeffrey McNeely, and Gonzalo Oviedo, 1–14. London: EarthScan.

Verschuuren, Bas, Robert Wild, Jeffrey McNeely, and Gonzalo Oviedo, eds. 2010. *Sacred Natural Sites, Conserving Culture and Nature.* London: EarthScan.

West, Paige. 2006. *Conservation Is Our Government Now: The Politics of Ecology in Papua New Guinea.* Durham, NC: Duke University Press.

White, Richard. 1991. *The Middle Ground: Indians, Empires, and Republicans in the Great Lakes Region, 1650–1815.* New York: Cambridge University Press.

Wild, Robert and Christopher McLeod. 2008. *Sacred Natural Sites. Guidelines for Protected Area Managers.* Best Practice Protected Area Guidelines Series, No. 16. Gland, Switzerland: IUCN; Paris: UNESCO.

Woodley, Steven, Bastian Bertzky, Nigel Crawhall, Nigel Dudley, Julia Miranda Londoño, Kathy MacKinnon, Ken Redford, and Trevor Sandwith. 2012. "Meeting Aichi Target 11: What Does Success Look Like for Protected Area Systems?" *PARKS* 18, no. 1: 23–36.

Chapter 2

Structural Changes in Latin American Spirituality

An Essay on the Geography of Religions

Axel Borsdorf

Introduction: The Religious Structure of Contemporary Latin America

According to figures given in the CIA factbook (CIA 2013) and the *Pontifical Yearbook* (Secretary of the State 2013), 80 percent of Latin American citizens are baptized in the Catholic rite; collectively, they account for more than half of all Catholics in the world. No other major cultural region has a higher percentage of Catholics. The visits of Pope John Paul II and Benedict XVI to the Latin American nations, including Cuba, show the prime attention the Vatican has paid to the largest Catholic region in the world. The first Pope born in Latin America, Francis, made a highly anticipated visit to his home region in July 2015.

Borsdorf (2013) has previously described the development of, and the different perspectives on, the geography of religions. With regard to Latin America, there have been many studies on religions, focusing on Catholicism (Berryman 1987; Coleman 1958; Rottländer 1992; Tworuschka 1996), the syncretistic cults (Métraux 1951; Steger 1970, 1987; Bastide 1978; Bramly 1978; Davis 1986; Coelsch 1989; Voeks 1993), or the Protestant denominations (Glazier 1980; Kohlhepp 1980; Martin 1993; Loth 1991; Brusco 1995; Palmié 1996). Borsdorf (1999) gave an overview of Latin American religions, and Borsdorf and Stadel (2015) provided a detailed analysis on Andean spirituality, in which they explore the notion of *lo andino,* the spiritual beliefs of the autochthonous Andean population.

In this chapter, a special emphasis on sacred sites adds to the understanding of the spatialities of indigenous revival encapsulated by the concept of *lo andino* by providing a framework for the dynamic changes in religious affiliations and spirituality in the region, especially those changes related to the reaffirmation of syncretic and/or original peoples' beliefs.

Latin American Catholicism is not a homogeneous entity; the affirmation of the Catholic theologian Rottländer (1992: 81) and the statements of Adveniat (2014), which defined Latin America as a Catholic continent and the Latin American Catholic Church as a homogeneous hegemonic force, must be criticized. It is important to note that most ethnic groups are not fully represented in census data and that the collection of survey questionnaires rarely include isolated indigenous territories. These surveys also do not consider whether people who have been baptized in the Catholic faith attend Mass on Sundays. Pizzey (1988) presented the case that Spanish priests performed a mass baptism of about forty thousand Indians at once; however, these people may or may not consider themselves Catholics.

While the European Catholic Church is more homogenous, in Latin America, Catholicism is differentiated into four main currents: formal Catholicism, nominal Catholicism, Liberation Theology, and syncretism. To the European visitor, Masses in many shrines do not represent the image of a Catholic continent. Oftentimes, monumental churches remain half-empty. According to some estimates, only between 10 and 15 percent of Catholics participate in Mass, and even fewer participate in the sacraments. Only a minority of those professing the Catholic faith in the state's census partakes in baptism, confirmation, and matrimony. One reason is extreme poverty; many citizens cannot afford to pay for the Church services. Thus, ecclesial matrimony is not the rule, and informal modes of coexistence are more common. In almost all countries, illegitimate births surpass the legitimate ones. Hence, for many Latin Americans, the Church functions as an institution whose rites are reserved for the high and middle classes of the population.

However, in these societal strata, a certain anticlericalism is often evident. This phenomenon is shown in a very Latin American way: criticism against the priest, the Pope, and the Church is always possible without questioning the quality of the members of one's own church. Coleman (1958) defines this form of Catholicism as "nominal." In regions of nominal Catholicism, the liberal clubs, like Rotary and Lions, have a strong influence.

On the other hand, "formal" participation in the Catholic denomination is characterized by a more intense collaboration between believers and the Church. In more conservative regions, more people participate in Mass, respect the ecclesiastic dogmas, and highlight the spirit of sacrifice. This form of religious life is located in the mining districts, in the rural areas with large landholdings, and in some regions of high and rapid urbanization. The regions where formal Catholicism is practiced include the coastal zones of Central America, the central highlands of Mexico, the Andes in Colombia, the coastal oasis of Peru, the slopes of the Argentinean Andes (the Littoral and Missiones), and the Brazilian Northeast.

But it is important to note that in many of these regions, the popular syncretic religion is very extensive among the lower classes. Therefore, even in the regions of formal Catholicism, only about 20 percent of the population participates actively in the Catholic ceremonies. Only in the Colombian Andes is this percentage much higher. In these regions, the conservative upper class and the ordained laicism are very powerful. In this region, it is an honor to have a family member serving as a priest or a monk.

Liberation Theology is a very different strain of Catholicism; it was formed during the Second Episcopalian Latin American Conference, held in Medellín, Colombia in 1968. It coincided with the launching of a quasirevolutionary movement whose themes were defined before important representatives of the Church and other intellectuals, such as Dom Helder Camara, Ernesto Cardenal, and Leonardo Boff (Boff 1987; Hennelly 1990). The perspective of Liberation Theology comes from "the bottom up": the poor people. Its methodology is the triple sequence see-judge-act, and its philosophy is based on the idea that the liberation of a person and a society links conversion, social justice, and charity. This reencounter with the poor is like a virtual spiritual interview with Jesus Christ. Indeed, poverty was seen by Liberation Theologians as a collective sin, one that will affect all communities in addition to individual, smaller sins. As an instrument to interact with the poor, Liberation Theology has developed the concept of basic ecclesial communities.

These communities are considered a new way of living Christianity, in the neighborhoods or villages, realizing the communitarian model of the Church. Already in the year 1980, Brazil alone had some eighty thousand basic communities with more than three million members. Liberation Theology was often regarded by conservative Catholics as being close to leftist ideologies. The breakdown of communism at the beginning of the 1990s meant that for the basic communities, the number of parishioners decreased as a result of losing many members to charismatic movements and syncretic communities.

Syncretism is the fourth form of Catholicism in Latin America. In Haiti, about 70 percent of the population practices Voodoo, despite having 80 percent of the population registered as being affiliated with the Catholic Church (CIA 2013). One third of Brazilians practice rituals from Afro-Brazilian religions such as Umbanda and Candomblé. Syncretism is characterized by a mixture of elements from popular Catholicism (adoration of saints, cults for relicts, or talismans) with magical and mythical elements of African origin or spiritual cults of indigenous origin. Popular Catholicism mixed with indigenous elements spread mainly in Mexico, Guatemala, Ecuador, Peru, Bolivia, the Toba regions of Argentina, and the Mapuche territory in Chile.

In spite of the mixture, some indigenous people, mainly in the low-land Amazon region, maintain their traditional religions. There are also non-Christian religions in countries with high Asian immigration, as in some Caribbean islands, the countries of the Guyana, and some large neighborhoods of Lima. In addition, the Jewish religion is represented with a community of about one million members in Argentina, Uruguay, Brazil, Mexico, and Dominican Republic.

Liberation Theology and syncretism change the worship, themes, and goals of the Catholic Church from within. But at present, the influence of Protestantism, with its many denominations, has intensified. As in the United States, the presence of the Lutherans is felt in areas where there was extensive European influx in the nineteenth century, as in Brazil, Argentina, and Chile, and the Anglican Church retains influence in Belize, Jamaica, and Guyana. Furthermore, the Moravian Church has a following in Jamaica and Suriname. However, today the most influential branch of Protestantism consists of groups with charismatic and evangelical orientation and with denominations of North American origin. Among these groups, only the Pentecostal movement is of South American origin. It was born in 1911 with a Methodist community in Chile, and today it is the fastest growing movement among all the religions of Latin America (Tworuschka 1996: 161).

It is possible to contest the idea that Latin America is presently experiencing a deep spiritual transformation, one that changes not only the face of religion but also the entire culture. The processes are, as will be demonstrated below, contradictory. They show regional concentrations that divide Latin America into a fascinating mosaic of minds, not only of people.

Elements of Latin American Spirituality

Hispanic Catholicism and the Character of the Conquest

Changes in the religious climate of Latin America are evident only through a reflection on the origin of spirituality imported from Spain in the sixteenth century. It must be remembered that in Spain, the elements of Arianism in early Christianity survived because the Roman influence under the dominion of the Moors was not strong. Some differences between Hispanic spirituality and Catholicism from the Vatican can be explained only from this perspective. In Arianism, Christ is not identical to God, but he moves toward God. He functions as a conduit to God, just as the Virgin Mary and the saints acted in other forms of Christianity. In Latin America, three significant cultic elements remain: the veneration of the saints, the widespread (in every country) worship of Virgin Mary, and the adoration of relicts. These three elements are central in Latin American Catholicism and are characteristic of popular devotion.

The Roman development of an episcopal Church was not relevant to Spain. Whereas the emperor of the Holy Roman Empire of the German Nation had to finally accept the Pope's central power in the Church, the Spanish king kept the legal power to name the archbishops, and he decided what religious orders the New World would have (Pérez 1995: 170). Furthermore, the Spanish Inquisition on the peninsula and in the colonies was not an ecclesiastic inquisition but a function of state under the king's control. The crown was the protectorate for the faith, but the king and queen also acted to control the cleric. There was a complete harmony between state and ecclesiastic interests. Thus, not only the missions to the Indies but also the Inquisition were tasks of the state, based on the first goal of the House of Austria, the creation of a World Catholic Union (see Borsdorf 1997).

The state's responsibility for religion was reflected in the institution of the encomienda, whereby the encomendero, as a representative of the state and of the Church, was charged with Christianizing the indigenous people. From this root, the declaration of the landowner, or hacendado, as lord, or señor, has a transcendental meaning. In the colonial capitalist system, from the revenues of the hacienda, the lord provided Indians with food and shelter with revenues from the hacienda, with the understanding that the lord was the sole owner of the landholding.

Given this connection, it is very important to understand that after the penetration of Islam into the Iberian peninsula and the Arab countries of North Africa, elements of Visigothic Arianism survived in Spain, Portugal, and the southern parts of Islamic territory: Eritrea and the Fon regions of Western Africa. Early Christian elements of so-called Augustinism from the Eastern Orthodox Church have survived in Ethiopia. This spread of Islam, which hindered the Roman Catholic influence, was demonstrated first by Frobenius (1933) and then by Steger (1987). From the two African regions, the Fon and Eritrea areas, originated spiritual currents for Latin America that found fertile ground for their Arianistic ideas.

Voodoo, Candomblé, and Umbanda developed from the spiritual elements of the Yoruba people that were not fundamentally different from Latin American animism. Rastafarianism was fertilized by religious elements from Ethiopia. In addition, the earlier Christian interpretation of the Holy Spirit as feminine (of which the Hagia Sophia in Istanbul is an example) finds its correspondence in the conception of Pentecostalism.

Depictions of a Black Madonna have a long history in Europe and are thought to be an influence of the Crusades. In Latin America they reflect the biblical adage *Nigra sum sed formosa* (I am dark, but beautiful) and, even more, indigenous and African American spirituality. Well-known examples of Black Madonnas are Nossa Senhora da Conceição in Aparecida, Brazil; La Virgen Morena in Andacollo, Chile; Nuestra Señora de Los Ángeles, Costa Rica; and La Divina Pastora in Siparia, Trinidad and Tobago.

Devotion to the Virgin Mary has a strong syncretistic element. The Aztecs also respected a virgin mother of God; in Peru and in Guatemala the indigenous people venerated, with Pachamama or with Ixchel, a similar feminine figure. The appearance in Mexico in 1531 of a dark-skinned virgin to the young Indian Juan Diego on top of the Hill of Tepeyac was the onset of one of the most powerful traditions of adoration in the Catholic world—the tradition of adoration of the divine indigenous women that continues today. The temple of the Virgen Morena, or the Guadalupana, in Mexico City is the main Catholic sanctuary of the whole of Latin America. A familiar element in the skyline of many cities of the subcontinent is a statue on the nearby hills of the virgin who blesses the citizens below. One emblematic example is the "dancing Virgin of Quito" in Ecuador, a giant effigy of the Immaculate Conception monument that towers on top of Panecillo hill, or Yavirak, following the design of Bernardo de Legarda, one of the master painters of the famed Quito's colonial school of art.

Every country contains such sanctuaries. Mamita Virgen leaves her mark on the daily life of the people. Images of the virgin are everywhere. Her mother-like being is similar to the image of Pachamama, and this is one way to understand her popularity.

Indigenous Religiosity and Indo-Catholic Syncretism

Despite differences between indigenous and Christian religions, there is some fundamental parallelism. The autochthonous religion was polytheist: the sun, moon, stars, rain, wind, soil, forest, and water bodies were gods, or at least abodes of the gods. The veneration of different saints, the cults devoted to Virgin Mary and Jesus Christ, and the adoration of relics facilitated the adoption of the Christian faith by the indigenous people.

Another important factor in the indigenous adoption of the Catholic faith was the fact that gods of nations that lost battles were tolerated as secondary gods, while gods of nations that were triumphant in battle were accorded respect. Religiosity was not as much the search for the truth as it is in Islam and Christianity. Thus, when Catholic missions appealed to the indigenous people, missionaries often did not insist on observing all traditional rites but only instituted the Sunday Mass (Clawson 1997: 192). While the Catholic faith was successfully transmitted in colonial cities, the indigenous mythology survived in rural areas and the periphery, where a lack of priests and churches only exacerbated the lack of enforcement of the mission by the encomenderos.

Some Catholic orders were also responsible for Indo-Catholic syncretism. For example, based on their contact with the indigenous people, the Dominicans acted with much tolerance; the Jesuits, along with the Dominicans, saw the Indians as sons of gods, tending more to the divine synthesis than to the radical persecution of heretic divergences.

In this interpretation, the precolonial gods—the Aztec Huitzilopochtli, the pre-Incan Viracocha, and the Mayan Kukulkán or Quetzalcoatl—mutated to the figure of Christ. The Aztec goddess Tonantzín was adored as the Virgin Mary (many faithful at Guadalupe worship Tonantzín, represented by the Virgin), and the saints replaced the pantheon of all relative gods. In Peru, the Catholic priests named Christ the "young lord of the Sun," and since 1944 they have honored him in an annual ceremony, known as the Inti-raymi festival (Métraux 1969: 191). Parallel to the custom during the mission of Germanic pagans in Europe, the Christian holidays were observed on the same day as the indigenous holidays.

Furthermore, in another parallel to the European mission, the sacred sites were also taken in the transfer of cultures: the church of Santo Domingo was built on top of the foundations of the Inka (also spelled Inca) temple of the sun, or Inti; and in Cholula, near the city of Puebla in Mexico, the largest pyramid in the world was crowned by a Christian church, as can be seen in Figure 2.1. Many Latin American cities are crowned by a huge figure of Jesus Christ, and cordillera passes are given his name. Rituals of sacrifices are in some cities realized at the entrance of the church, as in Chichicastenago, Guatemala, where the believers sacrifice corn or even chickens before they worship.

Finally, another parallel exists in the rites of sacrifices. Although the Church forbids human sacrifices, sacrifices for the Church are always welcome. As in Amerindian religions, these sacrifices exert influence on God's mercy. Nevertheless, today it is significant that popular Catholicism in the form of Indo-Catholic syncretism suffers a crisis equal to the variants of the formal Catholicism, nominal Catholicism, and Liberation Theology.

There are two tendencies responsible for this situation. First, the harmony of interests between representatives of the Church and holders of structural power motivated many poor people to abandon the official ceremonies of the Church. The community of Chamula in Mexico provides

Figure 2.1. Cholula, Mexico, church on the top of a pyramid (photo credit: Fausto Sarmiento)

a significant example. Here, all Christian elements of the Church were removed and replaced by the indigenous icons of the local indigenous religion. The other tendency is the progressive change toward a modern lifestyle, modeled on North America, and the modernization of even the peripheral areas. Literacy campaigns, growing mobility facilitated by modern communication routes, and modern medicine have produced a

Chololización indígena by making most mestizo and indigenous people a collective of urban dwellers. There, in the shantytowns or the marginal settlements where misery reigns, neither the shaman nor the priest plays an important role any longer.

However, some syncretic movements have experienced admirable growth. An example is the cult of María Lionza, which is widely spread throughout Venezuela (Canals i Vilageliu 2011). Followers of this mythology are understood to be Catholics, but in their understanding, God is far from humanity, and the people need ways to communicate with the divine force. Secondary gods establish this communication via the trance of a medium. Three gods form a court, a new trinity: María Lionza, a beautiful white woman; Guaicaipuro, an Indian god; and Black Felipe, an African American god.

Thus, the Marialionzistic cult operates as a bridge between the three ethnic cultures of Latin America and enjoys great popularity today, not only in Venezuela but also in the Caribbean islands and even in the United States. The sacred site of this religion is the Cerro de Sorte, located about 300 kilometers from Caracas. The most important ceremony takes place there on 12 October each year. The fundamental system of the mythology is similar to the cults of Voodoo and Candomblé, with origins in Arianism, native American cults, and African cults.

Afro-American Cults

Like the María Lionza movement, some African American syncretic cults have had an effect even among people of European ancestry. The Afro-Brazilian Umbanda is one of the most dynamic cults, as is the Cuban Santeria, which has experienced growth even in the United States (Palmié 1996), and not only among African Americans. The spiritual and esoteric elements of these cults satisfy the emotional needs of many people. However, because of their link with the feared militia Toton Macoute of the Duvalier family in Haiti and their close link to Voodoo, these cults have lost a major part of their influence in Haiti.

There is much variety among the African American cults. This is not surprising given the relationships among the different African religions in areas from which the slaves were brought. People taken involuntarily to the New World brought with them the religious ideas and myths of their homelands. But to become part of local society, they accommodated their spiritual beliefs to customary Catholic imagery. The easiest method was to build altars for Catholic saints where their traditional gods and goddesses could be venerated. In this manner, religion for the slaves changed to a more meditative affair that affirmed their cultural identification in the new environment, while expressing cultural resistance to the infiltration of white civilization. This function prevented African American religions from declining, as had happened to other Indo-American cults.

Due to the clandestine origin of these religions, the sacred places are widespread and have mostly only local importance. In Colombia, San Basilio de Palenque, a village that calls itself the first free black village of South America, may be seen as a sacred place for Afro-Colombians. In the center of the village, a statue of Benkos Biohó honors this leader of African slaves who led them to freedom.

This chapter cannot mention all the countries that have African American cults, but examples include (1) Brazil, where, in addition to Umbanda and Candomblé, there are Catimbó, Xangó, Calundu, Batuque, and Macumba; (2) Cuba, where there is Santeria with its different rules and cults, such as Palo Monte and Mayombe; also, Abakuá and Ñañigos; and (3) Suriname, where Ndyuka and Saramaka, or the cult Gaan Gadú, existed. In the syncretic religions of the Anglophone Caribbean, the Protestant influence is greater in the Shango and the Kele. In Jamaica, the revivalist tradition developed Rastafarianism, and in Trinidad and Tobago, Spiritual Baptists.

The cults of Voodoo and Rastafarianism will be examined as further examples of the African signature in Latin America religions. The term "Voodoo" comes from the Fon language of Dahoney and Benin and means "god." This god, very similar to the Christian god, is called "Bon Dieu." There are many intermediate gods, the loas, and there are many structural similarities to the María Lionza cult. The rites of obsession and sacrifice play an important role in Voodoo and Rastafarian religious life. In ceremonies, the loas take possession of the pious beginners—thus they take personification in reality. There are two different rites: *Rada,* the communion with the gods of Fon that are peaceful and favorable, and *Petro,* whereby the hostile loas bring disaster and grief.

There is a strong connection between Voodoo and Catholicism. The host from Mass, which plays an important role in the Voodoo cult, has to be blessed by a Catholic priest. The *hungan,* the Voodoo priest, has to partake in Holy Communion to steal one host for his rite. He also needs to bring holy water from the Catholic Church to his Voodoo temple. A person who is not a Catholic cannot be Voodooist; breaking with the Catholic Church means breaking with Voodoo. The person who converted to a Protestant religion not only terminated his or her syncretic religiosity but also cut ties with his or her cultural tradition. Both Voodoo and Santeria have strong political components. However, this is more imminent in Rastafarianism.

Rastafarianism was born in Jamaica during the 1930s and was based on the teachings of Marcus Garvey (1887–1940), who considered himself the reincarnation of John the Baptist. He prophesized in 1929 the coronation of a black king in Africa and the liberation and repatriation of all African Americans, those with black skin, to the larger pan-African homeland. With the coronation of Ras (Duke) Tafari Makonnen as the king of kings in Ethiopia in 1930, the first part of the prophecy of the reincarnation of god as a black man was fulfilled. Ras Tafari adopted the name Haile Selassi, meaning "power of the trinity." In his name, the Jamaican revivalists were called Rastafarians. When the king Haile Selassi visited Jamaica in April of 1966, the Rastafarians enthusiastically celebrated the presence of god, who traced his genealogy to King Solomon and the queen of Saba (Loth 1991).

Rastafarianism has strong Protestant roots. The Bible and asceticism (such as forbidding alcohol and meat consumption) are the bases of Rastafarianism. Communication with god happens through a personal, nondogmatic study of the Bible, sometimes stimulated by the consumption of ganja (cannabis). The goal is to experience the divine ego in the personal ego. Therefore, oftentimes Rastafarians talk about themselves as "I and I."

Rastafarians founded new forms of language, music, lifestyle, and personal appearance. Their most distinctive symbol is that of the dreadlocks hairstyle, which represents the African symbol of a lion. Sometimes they hide the dreadlocks behind tricolored wool hats: red for the blood of slaves, yellow for the greatness and beauty of Africa, and green for the longing and hope of returning to Africa. Rastas attempt a "hola life" both holy and holistic. They recognize Christmas and the New Year as holidays as well as the birthday of Ras Tafari and the day of his coronation. The music of the Rastas, known as reggae, seeks to intensify the contact of humans with god. The best-known Rastafarian musician was Bob Marley (1945–1981), who was believed to be the reincarnation of Marcus Garvey. The political content of his lyrics is evident and transparent partially only to Rastafarians. The Rastas use their own language or "dread talk," which includes several unique and rare expressions. To them, to speak for the Rastas is a revelation of the soul, and this manifestation is not frivolous. Language is a ritual in itself.

The central message of Rastafarianism is the fight against Babylon to free the blacks and repatriate them to Africa. "Burn Down Babylon," a song by Bob Marley, describes Babylon as the world of the white man. Through time, Rastafarianism found many converts in the Caribbean, coastal Honduras and Limón, Costa Rica, and the United States, while in Jamaica it currently suffers decline.

An important difference from Voodoo and other African American cults is that Rastafarianism does not have a tradition of venerating saints and does not have roots in the Yoruba religion. It is a phenomenon of the twentieth century, while the adoption of the Ethiopian religion has roots in early Christianity. With Voodoo, political intention is the start, but it is specifically reciprocated by a reverse racism (Hansing 2006). Racism exists more in Protestant than Catholic areas, whereas Rastafarianism has engendered a feeling of superiority of blacks over whites among its adherents.

Protestantism in Latin America

Protestantism——unlike in the colonies of Great Britain and the Netherlands in the Caribbean—in Latin America was born only after the wars of independence around 1850s. It was a consequence of the European policy on immigration and the strategy of the liberal parties and regimes to create a Protestant work ethic (Clawson 1997: 206). However, this early Protestantism was not attractive to Latin Americans, and it was concentrated in foreign colonies of European merchants and settlers. Also the Mennonites, migrating in the 1920s and 1930s mainly to Mexico, Belize, Bolivia, and Paraguay, remained closed, insular communities where missionary attempts to penetrate were unsuccessful. For Protestants, sacred sites do not exist, and even the church building is regarded as only somewhat sacred and only during worship services.

A missionary movement of charismatic evangelicals began in Chile. There, in 1909–1911, a Methodist Pentecostal church originated a trend that gained influence and adepts among the poor urban neighborhoods of Santiago. Meanwhile, in addition to the Pentecostal movements, many sects of North American origin were also diffusing, such as the Adventists, Mormons (Church of Jesus Christ of Latter-day Saints), Nacarens, and Jehovah's Witnesses. The traditional Protestant churches, the Anglicans, and to a lesser extent the Lutherans as well, copied spiritual elements from these denominations, which intensified their influence on the poor masses.

The only autochthonous religion in Latin America among the Protestant denominations is Pentecostalism (Figure 2.2). Pentecostalism does not occur as only one church, centralized with a directive power; there is instead a disconcerting variety of autonomous communities and churches, which sometimes are called evangelical, Pentecostals, or Assemblies. The most well-known are Assembly of God, Brasil for Christ, Church of God, Pentecostal Church of Chile, and the Pentecostal Mission. In common, they all trust in the illumination of the Holy Spirit and the Pentecostal experience, which includes the practice of speaking in tongues. They enjoy the manifestation of God in the form of visions or dreams as well as curing the sick with prayers and the "laying on of hands." It is the small community size and its transparency that confer the Church's power and attractiveness.

Figure 2.2. The "Virgen de Legarda" decorates the skyline of the Yavirak hill, in the city of Quito, Ecuador.
This giant statue has now crowned ancient sacred sites of the Inka that built the temple of the moon, and previous Shyri shrines that overlooked the capital of the kingdom of Kitu. After the colonization from Spain, the name "panecillo" is used to refer to the hill where many cisterns and aqueducts were built to provide Quito with its colonial architecture that is considered by UNESCO a World Cultural Heritage Site. (Photo: Fausto Sarmiento)

Many Churches do not have more than three hundred members, while others have only between thirty and fifty devotees. In this way, it is possible to offer all members help and love, which results in an absence of alcoholism, a rise in morale, and the adoption of a new work ethic. Because the Bible is to Pentecostals the secure source of inspiration, every parishioner is literate so they can read it. They all pursue a formal education. In addition, because family planning is permitted, contraceptive methods allow for small families, in which the father has a very responsible role. Some authors interpret this Latin American faith as the antithesis of the "fatherless society." Machismo, the claim of men to be superior to women, is not found among Pentecostals. In addition, all brothers are considered equals. This notion elevates the dignity and the consciousness of the value of each person, which is the pillar for social and economic progress.

The possibility of vertical mobility is interpreted as a testament to divine mercy, an interpretation that is basically Protestant and that strengthens the faith of all the members of the community. The Pentecostal work ethic encourages business owners to offer employment to Protestant workers because they have gained a reputation as hard workers who are agile and, above all, very safe and loyal. Because of this, joblessness and extreme poverty is much less common among Pentecostals and other Protestants than among the Catholic lower classes.

The charismatic movement of Pentecostals is the reason the fundamentalist Christian religion has seen higher growth rates in Latin America. The cause is not only economical, as we have seen above, but also spiritual. Pentecostals are well adapted to the structures of developing countries and to one of the roots of Latin American culture: Arianism, with its neglect of the holy trinity. Hence, the intensification of the Pentecostal influence will drive the future of religion in Latin America. This is notwithstanding the recent selection of the first Latin American Pope, Papa Francisco, who has had the effect of evoking their grandparents' affiliation with and patriotic sentiment toward Catholic traditions among Catholic youth.

Because they give the impression of success, Pentecostals have become the subjects of popular research among sociologists and geographers. Progressive scientists note their advances in education and health, their fight against alcoholism, and their adaptation to democratic processes in community life—in decision making as well as in the phenomenon of vertical mobility (Willems 1967). Marxist researchers interpret the evangelical movements as a decline of the consciousness for class struggles and see Pentecostals and other Protestant denominations as instruments of capitalism (Lalive d'Epinay 1969). This contestation presents an opportunity for some Catholic authors, who use the criticism to argue that there is a conspiracy against the "only religion" (Tworuschka 1996: 162)—the Roman Catholic Church's medieval accusations of heresy against Arians and Cathars is repeated.

The dynamism of evangelical movements and the high level of participation in Mass, on one hand, and the relatively low levels of practicing Catholicism, on the other hand, demonstrates that, at present in many Latin American countries, there are as many practicing Protestants as active Catholics. Martin (1993: 50) reports that in Brazil as early as the 1980s the number of Protestant pastors exceeded the number of Catholic priests.

Religious Transformation as Potential for Development?

Latin American religiosity and the religious map of Latin America are both undergoing deep structural changes. Based solely on the findings of the official censuses of Latin America, the region appears to be a Catholic continent. However, it must be noted that Latin American Catholicism is very different from that of Europe. It has many more syncretic forms, and even in its more dogmatic forms, it is interpreted in several ways. There are representatives of a formal Catholicism, while others are happy with nominal expressions of their faith; both of these groups have patterns that differ from those of the adherents of Liberation Theology.

The reaffirmation as multinational republics in the constitutions of Bolivia and Ecuador included support for the maintenance of traditional and ancestral beliefs. In Ecuador, the Shuar people have retained their respect for the waterfalls coming down from the Sangay volcano as sacred sites. There are also examples of the Otavaleños reaffirming their identity, including initiation rites in the Peguchi waterfall in the Imbakucha watershed.

Subtracting from the official numbers of nominal Catholics those who are nonpracticing as well as participants in African American cults or Indo-American syncretic cults, Catholics are in the minority. In addition, the Catholic Church's pretensions to being the only representation of the correct teachings of God is fought by the sects of North American origin and the even more powerful Pentecostals. Figure 2.3 shows a general view of the religious landscape of Latin America.

On the other hand, the Christian faith is not lost on the continent. Since its origins, the Ibero-American culture has been characterized by a strong link with the mystic, magic, occultism, spirituality, and even ecstasy. Today, these needs are satisfied not only by the Catholic Church but also by other institutions and movements that, even in the most extreme cases, have Christian foundations. Traditional medicine, for instance, calls for

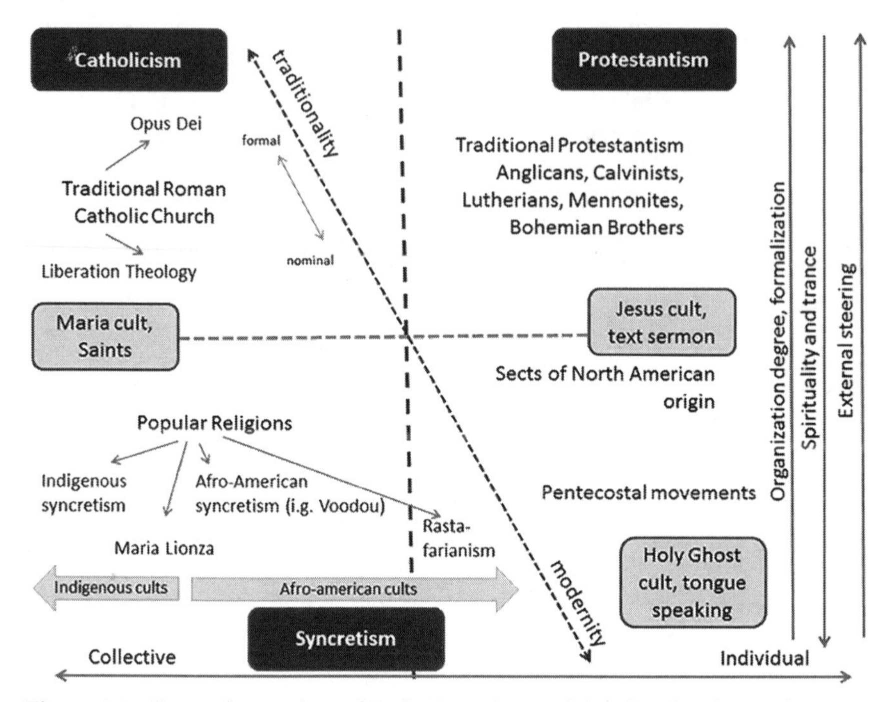

Figure 2.3. General overview of Latin American spirituality (credit: Axel Borsdorf)

sanitary practices or cleansing, often alluding to the power of the Catholic trinity to ward off negative influences, even evil possessions.

Throughout the continent, the Catholic affiliation with saints is brought to bear on the local scene of famous healers or makers of supernatural, miraculous treatments of disease; examples include Hermano Miguel in Ecuador, the Hermano Gregorio (José Miguel Gregorio) in Colombia, the Gauchito Gil (Figure 2.4) and the Difunta Correa in Argentina, and even the syncretic figure of Ekeko in Chile. Furthermore, the presence of New Age movement followers, mixing elements of Western and Eastern traditions in spirituality, confers on many capital cities the cosmopolitan choices of observance and respect.

To further complicate the spirituality mosaic in Latin America, there is a need to locate cults that have been created based on paranormal phenomena, such as the observation of UFOs (unidentified flying objects) and reported sites of extraterrestrial contacts, such in the lakes of Chilca, Peru, where a museum of the extraterrestrial experience welcomes visitors to this desert sacred highland. A similar trend is observed in Cerro Uritorco, in the locality of Capilla del Monte, Córdoba, Argentina, where hundreds,

Figure 2.4. Maria Lionza.
Maria Lionza is known as the "Queen" of the Venezuelan spiritual movement, a syncretic tendency that brings an indigenous goddess conceived as manifestation of beauty, peace and love, riding a tapir. The mountain of Sorte is the epicenter of reverence of pilgrims that visit the site for prayer and cleaning ceremonies every October. Ms. Tania Guzmán has spearheaded the initiative to request UNESCO recognition of the mystic Maria Lionza as an important cultural immaterial heritage of Latin America. (Photo: courtesy of Tania Guzmán)

if not thousands, of followers converge toward the idea of spiritual connection with foreign visitors.

Some cults, like those who engage in ceremonies involving Maximón in Guatemala and Ecuador, are not to be placed within the schema of differential Catholicism given in Figure 2.3. Maximón, a persiflage of a rich U.S. American, with his cigar and his pockets filled with U.S. dollars, is another extension of Catholic cults; people ask Maximón for help with financial or family issues.

Conclusion

This overview of the manifold religious landscape of Latin America has shown that this major cultural region no longer can be regarded to be a homogenous Catholic subcontinent. From the beginning of the conquest, the Catholic Church developed regional characteristics, expressed in a more intensive adoration of the Virgin Mary, which in the course of time has also been reflected in Europe. The saints, constitutive of the Latin American Christian belief, became even more important due to the importation of the African religions brought by slaves. Because of this, the rites of Voodoo, brought by the Fon and Yoruba tribes, as well as similar rites, have been able to survive in a syncretic form and extend their importance even to young white believers.

On the other hand, Liberation Theology, with its care for the underprivileged, its political struggle for more just conditions, and its emphatic ceremonies in the worship of the basic ecclesiastical communities, has made the old Church newly attractive for idealistic people as well as for the rural poor. Thus, Catholic sacred sites—churches, Black Madonnas and Black Christs, monuments, places of pilgrimage—never lost their importance. The famous "religiosity of the people" can be observed in many places, including the many wayside crosses, shrines, artificial or natural grottos (mostly called *lourdes*). Some places have an interregional importance, first among them being the site of the witnessing of the Virgin of Guadalupe.

Meanwhile, the traditional beliefs of the indigenous people have survived for centuries in rural areas of Latin America. Nowadays, old sacred places of the Indios are newly esteemed, and festivals of the sun goddess are even tolerated by the Church. Sacred places are the city of Cuzco, the Island of the Sun in the Titicaca Lake, and many other locations of specific natural value. The spread of Protestant denominations has meant a sudden and effective intrusion into traditional beliefs. Not only religions of North American origin but even more so the rise of the Pentecostal movement have broken into the traditional spirituality, motivating their believers to an individual study of the Bible and leading them to self-responsibility, asceticism, and hard work. These denominations do not need magic places, sacred symbols, or sacrificial altars. Today they are much more global than the Catholic Church, which in Latin America developed a specific regional face.

Identities, like the indigenous or African American, are overwhelmed by the individual access to God, Jesus Christ, or the Holy Ghost. As a consequence, this has led to a loss of societal coherence. Capitalist and egoistic thinking is no longer restricted to some economic actors but has become inherent to those members of lower classes who follow the new Protestant denominations.

Whether this process of religious change affects both Ibero-American culture and the potential to develop in the region is contested. If one defines the development problems in Latin America as resulting from an inequality of power, unfair distribution of wealth, income inequality, a corrupt and poorly run administration, and a lack of work ethic and responsibility for the environment, it can be stated that the reform movements within the Catholic Church contribute to a needed change. But even more effective in the change process is Protestantism, with its fight against alcoholism and joblessness as well as its effectiveness in bringing about vertical mobility, which translates into social and economic improvement for its members. It does not tolerate corruption and hypocrisy; moreover, its values are contrary to machismo and individualism. Protestantism is based on ascetics and the instinct of saving, along with hard work and individual ambition. These are qualities that Galtung (1994) declares as foundational to positive economic development.

Latin America is not turning away from Christianity; however, the variety of belief systems has been remarkably expanded, and the traditional Roman Catholic Church has lost a great deal of its influence. We must wait and see whether Papa Franciscus, the first Latin American Pope, can give new life to his Church in Latin America.

Axel Borsdorf, Ph.D., is emeritus full professor of geography, University of Innsbruck, and has been the director of the Institute for Interdisciplinary Mountain Research, Austrian Academy of Sciences. He is an expert on Andean migration and urban studies, community development, and biosphere reserves. He has been president of the Austrian Geographical Society and vice-president of the Austrian Latin-American Institute and has been awarded honorary membership in the Austrian Long-Term Ecological Research Program and the Franz-von-Hauer Medal, the highest award in Austrian geography. He taught as a visiting professor in Chile, Mexico, Switzerland, Thailand, and the United States and is coauthor of the book *The Andes: A Geographical Portrait.*

References

Adveniat. 2014. "Kirche in Lateinamerika." Available at: http://www.adveniat.de/lateinamerika/kircheinlateinamerika.html (accessed 2 February 2014).

Bastide, Roger. 1978. *The African Religions of Brazil: Toward a Sociology of the Interpenetration of Civilizations.* Johns Hopkins Studies in Atlantic History & Culture. Baltimore: Johns Hopkins University Press.

Berryman, Phillip. 1987. *Liberation Theology. Essential Facts about the Revolutionary Movement in Latin America and Beyond.* Philadelphia: Temple University Press.

Boff, Leonardo. 1987. *Introducing Liberation Theology.* Maryknoll: Orbis Books.

Borsdorf, Axel. 1997. "Stadt-und Kulturlandschaftsentwicklung in Lateinamerika unter den Habsburgern." In *Geisteshaltung und Stadtgestaltung,* edited by Manfred Büttner, 197–218. Frankfort: LIT.

Borsdorf, Axel. 1999. "Estructuras y Cambios en la Espiritualidad Latinoamericana: Un Ensayo de la Geografía de las Religiones." In *Coloquio Internacional*

Geografía de las Religions, edited by Blanca A. Fritschy, 57–76. Santa Fé de la Vera Cruz: Universidad Católica de Santa Fé.
Borsdorf, Axel and Martin Kumlehn. 2013. "Geography of Religion." In *Religion Past and Present*, edited by Hans-Dieter Betz, Don S. Browning, Bernd Janowski, and Eberhard Jüngel, 353–354. Leiden: Brill.
Borsdorf, Axel and Christoph Stadel. 2015. *The Andes. A Geographical Portrait*. Heidelberg, Dordrecht, and London: Springer.
Bramly, Serge. 1978. *Macumba, die Magische Religion Brasiliens*. Freiburg im Breisgau: Bauer Hermann Verlag.
Brusco, Elizabeth E. 1995. *The Reformation of Machismo: Evangelical Conversion and Gender in Colombia*. Austin: University of Texas Press.
Canals i Vilageliu, Roger. 2011. "Les Avatars du Regard dans le Culte à María Lionza (Venezuela)." *L'Homme* 198–199: 213–226.
Central Intelligence Agency [CIA]. 2013. "The World Factbook." Available at https://www.cia.gov/library/publications/the-world-factbook/fields/2122.html (accessed 2 February 2014).
Clawson, David L. 1997. *Latin America and the Caribbean: Lands and Peoples*. Dubuque: Oxford University Press.
Coelsch, Johannes. 1989. *Der Haitianische Vodoukult: Eine Religionsgeographische Studie*. Salzburg: Institut für Geographie.
Coleman, William Jackson. 1958. *Latin American Catholicism: A Self Evaluation*. Maryknoll: Maryknoll Publications.
Davis, E. Wade. 1986. *Die Toten kommen zurück. Die Erforschung der Voodoo-Kultur und ihrer geheimen Drogen*. Munich: Droemer-Knaur.
Deren, Maya. 1976 [1953]. *Divine Horsemen. The Living Gods of Haiti*. London: McPherson.
Frobenius, Leo. 2012 [1933]. *Kulturgeschichte Afrikas. Polegomena zu einer Historischen Gestaltlehre*. Zurich: Salzwasser Verlag.
Galtung, John. 1994. "Umbrüche im Norden—Verschärfung räumlicher Probleme im Süden." In *Verhandlungen des 49. Deutschen Geographentages Bochum 1993*, 10–22. Stuttgart: Steiner.
Glazier, Stephen D. 1980. *Perspectives on Pentecostalism: Case Studies from the Caribbean and Latin America*. Washington, D.C.: University Press of America.
Hansing, Kathrin. 2006. *Rasta, Race and Revolution*. Berlin: LIT.
Hennelly, Alfred T., ed. 1990. *Liberation Theology: A Documentary History*. Maryknoll: Orbis Books.
Hoover, Willis Collin. 1948. *Historia del Avivamiento Pentecostal en Chile*. Valparaíso: Imprenta Excelsior.
Kohlhepp, Gerd. 1980. "Bevölkerungs- und wirtschaftsgeographische Tendenzen in den mennonitischen Siedlungsgebieten des Chaco Boreal in Paraguay." *Tübinger Geographische Studien* 80: 367–405.
Lalive d'Epinay, Christian. 1969. *Haven of the Masses: A Study of the Pentecostal Movement in Chile*. London: Lutterworth.
Loth, Heinz-Jürgen. 1991. *Rastafar: Bibel und Afrikanische Spiritualität*. Cologne and Vienna: Böhlau.

Martin, David. 1993. *Tongues of Fire: The Explosion of Protestantism in Latin America.* Oxford: Wiley-Blackwell.
Métraux, Alfred. 1951. *Le Vaudou Haitien.* Paris: Gallimard.
Métraux, Alfred. 1969. *The History of the Incas.* New York: Random House.
Palmié, Stephan. 1996. "Afrokaribische Religionen." In *Religionen der Welt,* edited by Monika Tworuschka and Udo Tworuschka, 395–397. Munich: Bertelsmann Lexikon Verlag.
Pérez, Joseph. 1995. *Ferdinand und Isabella: Spanien zur Zeit der Katholischen Könige.* Munich: Callwey.
Pizzey, Jack. 1988. "Sweat of the Sun and Tears of the Moon: The South American Journey." Geographic essay for the BBC. London.
Rottländer, Peter. 1992. *Die Eroberung Amerikas und wir in Europa.* Aachen: Misereor.
Sawatzky, Harry Leonard. 1971. *They Sought a Country: Mennonite Colonization in Mexico.* Berkeley: University of California Press.
Secretary of the State. 2013. *Annuario Pontificio 2013* [Pontifical Yearbook 2013]. Vatican City: Secretary of the State.
Simpson, George Eaton. 1978. *Black Religions in the New World.* New York: Columbia University Press.
Steger, Hanns-Albert. 1970. "Revolutionäre Hintergründe des Kreolischen Synkretismus." *Internationales Jahrbuch für Religionssoziologie* 6: 99–139.
Steger, Hanns-Albert. 1987. "La Formación de la Conciencia Social y los Cultos 'Náñigo' en el Ámbito del Caribe Cubano." In *Kuba, Geschichte, Wirtschaft, Kultur,* edited by Titus Heydenreich, 9–24. Munich: Fink.
Tullis, Lamond. 1987. *Mormons in Mexico.* Logan: Utah State University Press.
Tworuschka, Monika and Udo Tworuschka, eds. 1996. *Religionen der Welt in Geschichte und Gegenwart.* Munich: Bertelsmann Lexikon Verlag.
Voeks, Robert A. 1993. "African Medicine and Magic in the Americas." *Geographical Review* 83: 66–78.
Willems, Emilio. 1967. *Followers of the New Faith: Culture Change and the Rise of Protestantism in Brazil and Chile.* Nashville: University of Illinois.

Part 2

Framing Sacred Sites in Indigenous Mindscapes

Introduction to Part 2

Framing Sacred Sites in Indigenous Mindscapes

Fausto Sarmiento and Sarah Hitchner

The chapters of this volume emphasize the ways that sacred natural sites have been framed in official conservation and development policies given the increasing acknowledgement of indigenous peoples' rights to self-identity and self-governance. Further, this volume showcases how various indigenous groups have initiated legal protections for sacred landscapes and natural sites as a means to not only protect the physical sites but also promote the preservation of their cultural heritage and identity, as discussed more generally in Part 1 (the Prologue and Chapters 1 and 2). To clarify and illustrate these two main points, we have separated the remaining chapters into two sections. Part 2 (Chapters 3 through 7) illuminates the panoptic perspective of the sacred dimension, or the ways in which sacrality is embedded in various landscapes and sites in the Americas as well as in international policies relating to sacred site conservation. Part 3 (Chapters 8 through 11) presents a series of case studies as specific examples of the trends identified herein.

As the theory of deep ecology (Naess 1973) gained traction, it evolved from a philosophical fad born of the American environmental movement of the 1970s; the theory focused on the intrinsic value and interconnectedness of all living beings and brought them into a rather applied social movement. This movement seeks a robust, modern society in which a collective ecocentric consciousness explicitly recognizes spiritual as well as physical connections among species (Naess and Kumar 1992). This trend has been criticized by some religious fundamentalists, who affirm that sanctity cannot be grasped by humans but can only emanate from a deity (Barnhill and Gottlieb 2010). The consolidation of a new field, so-called sacred ecology, appeared to provide academic grounding for the idea that "primitive" societies have environmental cognition—a cognition that often eludes "complex" Western societies (Berkes 1999).

Ecocritical rhetoric on poststructuralism and postmodern explanations of the deep ecology movement have led to increased scholarly engagement with the essence of place, including advancements in human geography, physical geography, and geospatial techniques (Glotfelty and Fromm 1996). Within the field of geography, sacred ecology is now represented in several geography specialty groups (GSGs) of the Association of American Geographers (i.e., the Emotional GSG, the Bible GSG, the Environmental Perception and Behavioral GSG, the Ethnic GSG, the Religions and Belief Systems GSG, and the Indigenous Peoples GSG).

Within the domain of anthropology, the American Anthropological Association has increased the membership of many groups associated with the themes of place, consciousness, and indigeneity (i.e., the Anthropology and Environment Society, the Association of Indigenous Anthropologists, the Society for the Anthropology of Religion, the Society for the Anthropology of Consciousness, and the Society for Cultural Anthropology) as well as of interest groups focusing on tourism, nongovernmental organizations, and nonprofit organizations. Bringing together these growing academic subfields, we include in Part 2 chapters (Chapters 3, 6, and 7) dealing with policies and management strategies for the preservation of places of cultural and spiritual significance as well as chapters (Chapters 4 and 5) that focus on how ecoregional settings, particularly sacred mountains, play an important role in the changing images and dimensions of indigenous identities.

References

Barnhill, David Landis and Roger S. Gottlieb, eds. 2010. *Deep Ecology and World Religions: New Essays on Sacred Ground.* Albany: State University of New York Press.

Berkes, Fikret. 1999. *Sacred Ecology: Traditional Ecological Knowledge and Resource Management.* Philadelphia: Taylor and Francis.

Glotfelty, Cheryll and Harold Fromm, eds. 1996. *The Ecocriticism Reader: Landmarks in Literary Ecology.* Athens: University of Georgia Press.

Naess, Arne. 1973. "The Shallow and the Deep, Long-Range Ecology Movement." *Inquiry* 16, nos. 1–4: 95–100.

Naess, Arne and Satish Kumar. 1992. *Deep Ecology* [film]. London: Phil Shepherd Production.

Chapter 3

El Buen Vivir and "the Good Life"

A South–North Binary Perspective on the Indigenous, the Sacred, and Their Conservation

Esmeralda Guevara and Larry M. Frolich

Introduction

Who is indigenous? What is sacred? Why should a particular site be conserved? These are questions that are inevitably framed, and answered when possible, within the cultural milieu of the individuals involved. Interestingly, traditional political and social divisions do not easily predict how people or groups of people might respond to these questions. We suggest that, at least for the Americas, a binary system of analysis that contrasts a technologically savvy, materials-rich, Global North perspective with an artisanship savvy, spiritually rich, Global South perspective might be the best frame for analyzing the notion of sacred, understanding how decisions about conservation are reached, and even knowing who is indigenous.

We characterize the Global South view with the notion of *el Buen Vivir,* as recently codified in the constitutions of Bolivia and Ecuador, and contrast it with the so-called Good Life, ironically a direct literal translation into English but practically a conceptual opposite. The very fact that these terms are literal language translations but conceptual opposites suggests that they might be better understood not as a dichotomy but as a binary system in which they as much depend upon, as contradict, each other.

We further make the case that to a very large extent, but not completely, the Global South–Global North binary is part of a demographic continuum, from *low* to *high,* of resource and power use that has accompanied urbanization and has led to a dramatic transformation in human lifestyles and human ecologies. This grand and fundamental conversion of the human bioscape has been transpiring over the last two centuries, initiating in the Global North, and it will continue well into the present century as it spreads thoroughly into the Global South.

Whereas the demographic/power use/urbanization transition provides a linear transect upon which some of the Global South–Global North contrast fits, we believe that, especially when considering the cultural environment that has fostered notions of *el Buen Vivir* and "the Good Life," the binary is more complex. Our hope is that a full exposition of the South–North binary, and how it envelops our current demographic and lifestyle transition, might offer a robust way of contemplating and deeply understanding notions of the sacred, the indigenous, and their conservation. We offer, as an extended example, an explanation of what the binary tells us about the quintessential environmental issue of population growth, arguably an essential element of what indigeneity and sacred mean to a particular human culture.

El Buen Vivir and the Good Life

Any time we endeavor to characterize or generalize about a particular culture, we run the risk of creating a parody; the larger the cultural, demographic, or geographic group that we wish to characterize, the greater the risk. With this caveat in mind, we believe that something essential, deeply rooted, and inherent to the cultural, linguistic, and political history of hemispheric South versus hemispheric North in the Americas has led to some distinct differences in how people view themselves and what motivations guide the way they live their lives.

Recognizing that the "fit" is never perfect, that people are as diverse as our imaginations will allow, and that each human life is a universe unto itself, we are willing to risk observing that something fundamental can be captured about the cultural milieu of the Global South by the notion of *el Buen Vivir* and that this can be contrasted with something fundamental about the Global North and what it means to live the Good Life. Less than trying to categorize or pigeonhole people, we hope to bring out the significance of language and how its use might guide the way our thinking is enabled or constrained.

At the same time, we recognize that we can no longer understand the North–South divide as strictly geographic. The notion of an isolated "Third World" made up of entire countries and regions that are impoverished has long been false, as globalization and urbanization bring pockets of high economic wealth to "underdeveloped" countries (Gilbert 1993; Varela 1998), while at the same time, many traditionally "First World" countries have seen their most economically impoverished populations overlap with Third World levels of economic poverty (Ariza 2003; Rosling 2009; Kneebone and Garr 2010).

Thus, North and South worldviews overlap both geographically and culturally (Sarmiento 2003), principally through economic migration that brings Global South views into the United States, as well as through the exportation of cultural works (TV, cinema, and music) and propaganda that create a Global North mindset among wealthy Southern urban dwellers (Gordon et al. 2010). With this in mind, we recognize that the terms "North" and "South," insomuch as they might have their roots in a real geographic concept, better refer to a mindset and lifestyle, and we must recognize that even the mindset is no longer geographically isolated.

As Lewis and Wigen (1997) have so elegantly pointed out, even the way we divide our planet into continents, although presented as a physical reality, is in fact a cultural construct. These authors bring a profound historical view to how notions of East, West, North, and South have developed, finally arguing for reconceptualizing Europe as Western Eurasia and the Americas as a continent that is culturally and physically divided into North America, Ibero-America, and African-America. In many respects, our conceptualization of "South" is in large part what they would consider Ibero-America, although culturally many, especially wealthy, Ibero-Americans and many who live in the geographically northern United States and Canada might now be considered part of North America.

At the same time, some populations in the southerly geographic Ibero-America, again especially the wealthiest parts, might be culturally more comfortable as part of what Lewis and Wigen consider "North America." So the divisions between North and South are not clear-cut, geographically or culturally, and it is best to accept them as diffuse and constantly changing concepts. A full application of the North–South binary to ideas about the indigenous revival and sacred site conservation should include a robust treatment of the African-America geographic and cultural framework, something that is beyond the scope of this chapter.

Keeping in mind the risks involved in generalizing about a large group of people, and recognizing that South and North are isolated neither geographically nor culturally, let us examine to what extent we can understand how *el Buen Vivir* has grown out of the Global South and the Good Life has come to symbolize the Global North.

El Buen Vivir *and the Global South*

El Buen Vivir, as a relatively new catchphrase in the Global South, has emerged from the Andean, Quechua/Incan-based concept of *sumak kawsay,* which suggests an ongoing relationship with the *pachamama,* which is, in turn, what might be called "the Earth" or "the soil" when viewed as a maternal, life-giving force ("Mother Earth" with all its political and social loading might not be the best translation here). Given this broad origin for *el Buen Vivir,* we cannot assume a universal definition or understanding for its significance in the larger Global South.

However, some common characteristics, in what is a multidimensional concept, have been identified, in particular by Gudynas and Acosta (2012). They view the concept of *el Buen Vivir* as a rejection of classic economic development in which the principal goal is enhancing the resource consumption and monetary standards of a country, especially as they are indicated by GDP (gross domestic product) and unemployment rates. *El Buen Vivir* does not accept that this economic view of so-called advancement or development is what necessarily leads to a better quality of life for the people of a region. This rejection also brings into question the colonial mindset, with its reliance on outside forces for change and its assumption that the colonizers, whether from the American North or Europe, are somehow better equipped, more knowledgeable, and live a higher quality of life that should be emulated.

An ethical component of *el Buen Vivir* has also been invoked (Gudynas and Acosta 2012). This includes recognition of an intrinsic value for nature, which has been codified in the new Ecuadorian constitution by granting fundamental rights to the natural world (Cardenas 2010). However, distinct from the traditional Northern environmental movement, the intrinsic values in nature are seen to bring a different ethos to how people live their lives, an ethos that lifts and beholds human quality of life by bringing noneconomic "nature-based" factors like spirituality, family, creativity, and relationships into the full picture of what it means to live a fulfilled life. This is in sharp contrast to the environmentalist perspective in which the intrinsic value of nature means that people must sacrifice something from their Good Life in order to defend or permit the conservation of nature, an idea that will be more fully explored later in the chapter.

The economic and ethical components of *el Buen Vivir* are not viewed as being in conflict, or as somehow working toward opposite goals (a common view regarding the Good Life in the North), but rather as significantly overlapping and synergistic contributions to overall human quality of life. One, now well-rooted, expression of this total picture of quality of life comes from the work of Chilean economist Max-Neef (1992) who, for many decades now, has promoted a multifactor analysis of life fulfillment. His "nine basic needs"—subsistence, protection, affection, understanding, participation, leisure, creation, identity, and freedom—are not ranked hierarchically, are not considered to be in conflict, and include combined economic and ethical–spiritual components (Table 3.1).

Table 3.1. Taxonomy of fundamental human needs (adapted from Max-Neef 1992: 32–33)
Max-Neef believes that the extent to which these needs are being met, in all the different ways of meeting them, should be the true measure of the wealth of a nation, society, or population.

Need	Being (qualities)	Having (things)	Doing (actions)	Interacting (settings)
Subsistence	physical and mental health	food, shelter, work	feed, clothe, rest, work	living environment, social setting
Protection	care, adaptability, autonomy	social security, health systems, work	cooperate, plan, take care of, help	social environment, dwelling
Affection	respect, sense of humor, generosity, sensuality	friendships, family, relationships with nature	share, take care of, make love, express emotions	privacy, intimate spaces of togetherness
Understanding	critical capacity, curiosity, intuition	literature, teachers, policies, educational	analyze, study, meditate, investigate	schools, families, univer-sities, communities
Participation	receptiveness, dedication, sense of humor	responsibilities, duties, work, rights	cooperate, dissent, express opinions	associations, parties, churches, neighborhoods
Leisure	imagination, tranquility, spontaneity	games, parties, peace of mind	daydream, remember, relax, have fun	landscapes, intimate spaces, places to be alone
Creation	imagination, boldness, inventiveness, curiosity	abilities, skills, work, techniques	invent, build, design, work, compose, interpret	spaces for expression, workshops, audiences
Identity	sense of belonging, self-esteem, consistency	language, religions, work, customs, values, norms	get to know oneself, grow, commit oneself	places one belongs to, everyday settings
Freedom	autonomy, passion, self-esteem, open-mindedness	equal rights	dissent, choose, run risks, develop awareness	anywhere

The combined ethical–economic view of development that arises from *el Buen Vivir* leads to very different goals and conclusions regarding not only the economic development of the country and the social structure of society but also what is considered sacred, how it is incorporated into the larger social milieu, and how conservation, if it is applicable, is carried out. The role of indigenous people in society, as a result, emerges in a very different, and generally much more dominant way when a *Buen Vivir* mindset is predominant. All this contrasts significantly with how the notion of the Good Life guides the Global North.

The Good Life and the Global North

Whereas *el Buen Vivir* has fairly recent and political–academic etymological roots (although the Incan/Quichuan roots of the concept are ancient), the Good Life is mostly a popular culture concept that emerged during the heyday of U.S. advertising and consumerism in the 1950s and 1960s (Belk and Pollay 1985). Since this era, the concept of the Good Life has infiltrated every aspect of Global North culture; in its mainstream interpretation, it is an exclusively economic or financial concept that involves obtaining all of the material wealth and possessions typically identified—explicitly through advertising or implicitly by the cultural milieu as forming part of the Good Life.

The idea permeates all aspects of Global North culture as indicated by this Wikipedia "disambiguation" page (http://en.wikipedia.org/wiki/The_Good_Life).[1] Sometimes the popular culture, expressing the deep and intransient roots of the economic basis for the Good Life, explicitly pushes back against the fundamentally economic measure of a fulfilled life. In the 1970s British sitcom *The Good Life* (later remade for American TV as *Good Neighbors*), the focus family is a financially independent couple who quit their regular work to establish a fully self-sustaining household on a typical suburban street, with the requisite counterculture gardens, farm animals, and methane-based electric production. This kind of deconstruction of the Good Life helps reiterate its fundamentally economic basis.

At a political and social level, the ingrained acceptance of the Good Life as derived from the financial well-being of the household means that societies and countries are also measured and compared on the basis of their economic statistics, GDP, and unemployment rates. Whereas these numbers might be good indicators of the relative political power that a country or society carries, it becomes increasingly evident that they do not perfectly represent the well-being, happiness, or quality of life that individual members of society might experience. Increasingly, alternative measures of life fulfillment, quality, or happiness are being developed (see Table 3.2), and when applied they often contradict the dominance order that arises from using GDP as a measure of countries' well-being.

Even many mainstream economists recognize that GDP is limited in terms of what it measures. Specifically, nonmarket goods, such as ecosystem services, unpaid labor, and leisure are not included in GDP (Wesselink et al. 2007). Depending on how these parameters are defined, and what is included, they are potentially greater contributors to economic activity and power use cycling than the market-based goods that *are* included in GDP. Additionally, GDP focuses on the current economic flow, or the most recent levels of power use, that is, the cycling of energy during the ultimate measurement period (usually weekly or monthly measurements that are scaled up to a single year). Capital reserves—be they natural capital in the form of soil, mineral, water, and air resources; human capital in the potential for human work; or even economic capital in the form of savings and infrastructure—are not reflected in the GDP calculation.

Numerous metrics and indices have been developed to account for these shortcomings in GDP (see Table 3.2 for a partial list). This list, to which new metrics are continuously being added, points out how GDP fails to account for the real quality of life, or well-being, that people experience on a daily level.

In the traditional economic view in predominately Global North, Good Life countries, conservation of sacred and indigenous sites or lifestyles is often identified as a "social good" that runs contrary to the development and well-being of the country, as measured economically by GDP. The ability to undertake these sacrifices represents the economic power of the country, and since it is usually portrayed as an economic sacrifice at the social and individual level, such sacred, natural, and indigenous conservation projects are at risk of being eliminated when economic indicators fall.

For many years, the typical colonialist "development package" that the Global North has offered to (or imposed upon) the Global South has stipulated the implementation of Global North-style conservation projects, thereby ensuring that the sacred and indigenous social goods are incorporated into the entire development picture. In fact, and perhaps representing the frustration with directly "developing" the mainstream economy, such conservation projects are sometimes the keystones of foreign aid, since they can more easily be defined (and morally defended) from a Global North perspective; thus, they ensure adoption of some fundamental tenets of the Global North Good Life worldview (even if it is, ironically, the adoption of the secondary social good leftovers of this worldview).

In stark contrast to the Global South's *el Buen Vivir,* the Good Life perspective marginalizes the indigenous, the sacred, and the natural, or at

Table 3.2. Alternatives to GDP

This table shows a partial list of alternatives to GDP for measuring economic–ecological activity and its consequences for human well-being. A full analysis of all these metrics, their utility, and the extent to which they have been applied is beyond the scope of this chapter. A few of the more pertinent and enlightening examples are discussed in the text.

Name of Indicator	Developer of Index	How Differs from GDP	Source
Indicator of Sustainable Economic Welfare	Friends of the Earth	Includes economic inequalities in a number of natural and human environmental indicators	Friends of the Earth, n.d.
Wealth Estimates	World Bank	Includes produced, natural, and intangible capital	World Bank 2010
Adjusted Net Savings	World Bank	Includes true savings rate after investments in human capital and depletion of natural resources and pollution	World Bank 2013
Human Development Index (HDI) and related indices	United Nations Development Programme	Measures human achievement based on long and healthy life, access to knowledge, and decent standard of living	United Nations Development Programme 2011
Environmental Performance Index (EPI)	Yale University	Tracks policy categories related to environmental, public health, and ecosystem vitality	Yale University 2012
Environmental Sustainability Index (ESI)	Yale University	Composite tracking socioeconomic, environmental, and institutional indicators	Yale University 2005
Sustainable Society Index (SSI)	Sustainable Society Foundation (SSF)	Includes twenty-four categories related to human, environmental, and economic well-being	Sustainable Society Foundation 2010
Canadian Index of Wellbeing (CIW)	Canadian Index of Wellbeing Network	Not a single index but includes information in many categories of well-being: living standard, healthy populations, community vitality, democratic engagement, time use, leisure and culture, and education	International Institute for Sustainable Development 2012
Ecological Footprint	Global Footprint Network	Measures how fast resources are consumed and waste is generated compared to how fast new resources can be generated and waste absorbed	Global Footprint Network 2012
Gross National Happiness	Centre for Bhutan Studies	Incorporates psychological well-being, time use, community vitality, culture, health, education, environmental diversity, living standard, and governance	Centre for Bhutan Studies 2012
Happy Planet Index (HPI)	New Economics Foundation	(see text)	New Economics Foundation 2012

least views them as secondary to the focus of human activity: economic undertakings and the achievement of fulfillment or higher quality of life. Nonetheless, and perhaps because of this special place for the idea of conservation of the sacred and the natural, some of the resultant conservation projects—the U.S. National Park System would be a prime example—are truly impressive in their organization, extent, and power.

Before more deeply analyzing the contrasting binary significance that Global South and Global North bring to our ideas of the sacred, the indigenous, and their conservation, we explore how these binaries fit into an economic–ecological model for power use, especially as it relates to human quality of life.

Power Use and the South–North Binary

Although the *Buen Vivir* perspective argues for a new model of development that takes into account more than economic factors, it is easy to identify a demographic trend, dating back to the early nineteenth century, that involves greater resource use, higher levels of GDP, urbanization of the population, and, it could be argued, a general move toward the Good Life and the identification of higher life quality with better access to material goods.

The question then arises as to whether our South–North binary is false and represents nothing more than an inevitable trend toward economic development and higher levels of resource use as well as whether the binary is in fact nothing more than extremes on a power use transect. The answer, we believe, is, in part yes to both, because the binary is best understood as a trend toward increased power use, and in part no to both, because the binary encapsulates something larger about the cultures involved. To arrive at that answer, we see a need to reduce the economics (and ecologics) of the South–North binary to their essentials, in particular to how we humans, as biological entities, use power.

Intuitive Parsing of Power Use According to Energy Level and Time Frame

A typical sixty-kilogram (130-pound) human, in order to power the daily functioning, or metabolism, of his or her body, uses power at a rate of about 130 watts, a little bit more than the brightest incandescent household lightbulb. However, virtually all people of today's world use significantly greater amounts of power than what their body's metabolism consumes (or converts). This additional, what can be called "extrametabolic," power is supplied by environmental fuel sources (fuel woods, hydroelectric and wind power, nuclear energy, and, most importantly, fossil fuels).

In order to parse human power use in a way that brings together a combined economic–ecological perspective, we analyze the human biological use of power in terms of its physical (or economic) as well as spiritual (or noneconomic and creative) energetic components. But what is meant by spiritual energy, and can it be distinguished from physical energy?

Recall, from basic physical principles, that

$$\text{Power } (P) = \text{Energy } (E)/\text{Time } (t),$$

and that improved performance in a complex system is equivalent to an increase in power.

Vermeij (2004) argues that in economic systems, performance can be understood as profit, in ecosystems as net productivity, and in evolutionary systems as reproductive fitness. Furthermore, he argues that there is no substantive difference between the three measures of performance, except that in economic systems, money can be used as a placeholder for energy in the power equation. In general, economic, ecological, and evolutionary systems move toward increasing performance, although this is not a continuous upward trend but one that is marked by cycles of lower performance and sometimes even periods of catastrophic reduction in performance and complexity (as in Holling's 2001 concept of panarchies).

The basic power equation gives us two ways to increase power use. One is by increasing the amount of *energy* that is being cycled and the other is to decrease the *time* over which a given quantity of energy is being cycled. A robust analysis of power use in the performance of complex systems needs to consider both factors. Increases in the amount of energy that is being cycled usually accompany *growth* of the system, whereas decreases in the time cycle accompany improved *efficiency* of the system. Growth and efficiency are typically considered to be absolute positives or desirable outcomes of change in the system, especially for human economic systems (Vermeij 2004; McConnell and Brue 2005). Ecosystem analyses of power use also see advantage or superiority in systems that exhibit growth and efficiency (Fath et al. 2004). In the evolutionary analysis of reproductive fitness, a deliberate analysis of the time frame for reproduction, in terms of what are called r-selection and K-selection strategies, recognizes that there is no intrinsic advantage to a particular time frame or level of energy input into offspring.

The message from r/K analysis is that in order to fully understand changes in performance related to power use or energy cycling, we need to take into account both energy and time variables and recognize that the overall success or health of the system does not depend only on increased growth and efficiency. In order to accomplish this kind of complete analysis, we can create a graphical space where different realms of power use are represented as the intersection between the amount of energy that is cycled and the time period over which that energy is cycled (Figure 3.1).

In Figure 3.1, each quadrant intuitively represents a different realm of power use, or a different way of doing work. A given entity, or energy cycler, be it an individual organism such as a person, an ecosystem such as a forest, or an economic entity such as a business, cannot be in more than one realm or even one exact point in the power use space at any given time.

The key question that emerges, then, is what point in the power use realm is ideal for a given human activity, such as conserving sacred sites? Extremes are probably to be avoided. For example, cycling high levels of energy over very short time periods results in explosive devices that are very hard to control. Although some would argue that bombs have a very

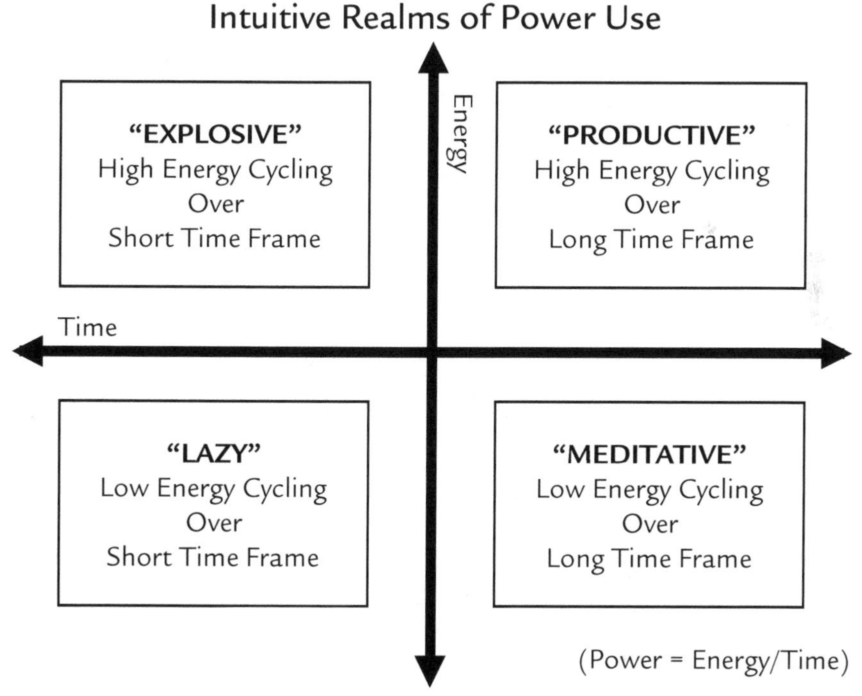

Figure 3.1. Intuitive realms of power use in a space represented by time and energy axes
This analysis is deliberately nonmathematical, but it should help to give an intuitive sense of how changes in energy use level over different relative time periods result in different types of power use regimes.

important role to play in certain circumstances, for most human activities they would be considered not only useless but highly destructive. The more energy that is cycled and the shorter the time period over which it is cycled, the more destructive the device. On the other hand, extremely low-energy cycling over very long periods of time probably reflects a non-living situation such as what we might find in inert solid materials.

For human individuals and groups/societies/economies, the use of extrametabolic energy (usually in the form of fossil fuels), often over relatively short time periods, tends to keep us in the productive or even explosive realms of power use. Intuitively, an argument could be made that spiritual energy cycling or power rarely involves extrametabolic energy and usually occurs over longer periods of time; such spiritually oriented activities—for example, conversation, family time, cooking from scratch, eating, prayer, meditation, communing with nature—would be found in the meditative or lazy realms. Among high extrametabolic use populations, specifically in the consumerist Global North, we hear frequent calls for slowing down to spend quality time in what we, at least intuitively, might consider a more spiritual cycling of energy. Below, we analyze whether the ideas of "indigenous" and "sacred" are more closely allied with lower levels of power use.

Are Lower, Metabolic-Level Power Uses Always Better, and Are They Always More Spiritual?

At the low-end power use extreme, the anthropological literature sometimes portrays hunter–gatherer, "uncontacted" peoples as having ideal lives with low stress and high levels of leisure time (Liedloff 1977). Although hunter–gatherers may have formed "the original affluent society" (Sahlins 1974), this way of life also carries the cost of high infant mortality, short lifespans, and the potential for periods of famine, direct violence, and disruption. Agricultural peoples who work the land with little or no extrametabolic input may have even higher levels of infant mortality and shorter lifespans, along with heavier workloads, frequent famine, and little time for any other activity other than work devoted to mere existence (Waters 2007).

At the other extreme on the scale of power use, we encounter the superwealthy, in economic terms. Here, anthropological studies are few, but the daily media presents us with a typical view of this lifestyle, in which children are often marginalized with little parental guidance and families are frequently disrupted or dysfunctional.

It may be that the Global South–North binary represents extremes on this transect from lower to higher power cycling and that we are simply defining where two geographic regions are, on average, at this point in history along the demographic transition toward higher energy use, lower fertility rates, and urbanization. However, we also see that cultural values accompany economics and resource use, and this may trump long-term demographic changes. A Global South perspective is qualitatively distinct from a Global North perspective, and the evidence is in how the South adapts to increasing power use at an individual, family, and population level.

In fact, *el Buen Vivir* has emerged in the very countries that are experiencing economic booms and a trend toward higher levels of resource and power use. And so despite an ongoing demographic and power use transition, the Global South, under a *Buen Vivir* perspective, demonstrates distinct differences in how individuals seek life fulfillment from more than economic means, how families maintain cohesiveness even as energy use and resources become more abundant, and how societies and political entities are more open to an economically distributive model that defines progress in more than financial terms. By the same token, economically poor populations that are culturally rooted in the Global North are likely to see their ultimate life satisfaction principally derived from the pursuit of material wealth or the Good Life.

A telling example of how demographics work with the cultural North–South binary involves the classic conservation–environmentalist issue of population growth. By applying our economic–ecological model for analyzing power use, data on fertility rates and power use reveal underlying cultural assumptions regarding how we understand this quintessential twentieth-century issue. Around the world, indigenous populations are typically characterized as having high birth rates, making this issue core to the question of how we define and determine indigeneity as well.

Population Increase: Do We Get to Decide How Many Children to Have?

Here we discuss how the South–North binary turns the ultimate conservation threat—population increase—on its head. James Brown, a theoretical ecologist at the University of New Mexico, has devoted his academic career to analyzing and compiling large amounts of ecological data and explaining those data based on size-related physiological constraints of the organisms involved. A fascinating analysis by Brown and his colleague Melanie Moses, a computer scientist with an interest in biological networks (Moses and Brown 2003), is worth reviewing in detail as a superb example of how power use analysis can give us insight into the Global South–Global North binary and its significance to the conservation of the sacred and indigenous.

And in a larger context, the power use analysis of Moses and Brown's data has implications for how we respond to the "limits to growth" or what affects the sustainability of a human economic–ecological system. In their article, Moses and Brown quantitatively analyze huge geographic and historical differences in human energy consumption along with changes in fertility rates using exponential scaling relationships based on body size.

For all mammals, energy use, or metabolic rate (B) is positively correlated with their size or body mass (M) under the well-developed, empirically supported exponential (or scaling) equation

$$B = B_0 \times M1/4, \text{ where } B_0 \text{ is a scaling constant.}$$

This equation predicts the observed typical metabolic rate for a sixty-kilogram human of approximately 2,500 calories per day, or 130 watts, a little bit more power use than a typical incandescent lightbulb.

For most organisms, total energy use is accounted for by the metabolic energy needed to fuel the body's cells, and so for a typical sixty-kilogram human, a 2,500-calorie daily diet should serve to fuel all the body's cells and thereby keep its total physiology functioning. In the case of humans, however, as we have already seen, technology takes us far beyond simple metabolic energy needs, and energy consumption includes not just food intake but also the consumption of oil, gas, electricity, and other energy resources, or what Moses and Brown (2003: 296) call "extrametabolic" energy. This extrametabolic energy consumption ranges from a few hundred watts per capita in the poorest countries to many thousands of watts in developed countries. In the United States, per capita energy consumption in the year 2000 was eleven thousand watts, which is about one hundred times the basic metabolic or food intake consumption. Moses and Brown point out that if this level of energy consumption is plugged into the energy-scaling relationship, the result is that the average modern U.S. citizen consumes energy as if he or she were a 30,000-kilogram primate.

According to Moses and Brown, other biological rates, such as cellular processes and reproductive rates, scale as body mass to the –1/4 power, and so, doing the math, it results that most biological rates (R) should scale with metabolic rate to the –1/3 power or

$$R \text{ is proportional to } B - 1/3$$

Since fertility rates are also well known for many mammals, the relationship between fertility rate and metabolic rate can be plotted, and the resultant scaling exponent closely matches the prediction from this second equation. Moses and Brown (2003) present summary data for mammals and primates to show the scaling of fertility rates among mammals and a time series from the United States to show fertility rates as power use increases in human populations (see Figure 3.1).

They then go on to compare the great range of human extrametabolic energy consumption with fertility rates. Their data come from over one hundred nations and span the time period from 1970 to 1997. They find that the very same predicted and observed –1/3 scaling exponent applies when comparing human energy consumption with fertility. In other words, just as for all mammals, overall metabolic rate (now including extrametabolic energy use) and the resultant virtual body size, can explain the nearly tenfold range of human fertility rates. Economically poor countries, where energy consumption is barely beyond typical food intake, have very high-fertility rates. However, in economically developed countries, where energy consumption is extremely high, fertility rates drop, as individuals extend their reach into the energy network to a point of consuming as if they weighed tens of thousands of kilograms.

This scaling relationship in humans reflects the basic trend for larger mammals to invest far more in their own metabolic upkeep and to have a small number of high-needs offspring. In the human case, among some developed nations, fertility is actually below two (which would be needed for replacement of the population), and the energy-extended current human population is actually shrinking.

From a Global North perspective (from which the Good Life promotes individual liberty in all aspects of our lives), we assume that a particular human behavior, like reproduction, reflects conscious decisions on the parts of the individuals involved. Demographic analyses of human population growth speak of the "demographic transition," from high fertility to low fertility, which inevitably accompanies increased economic wealth (Caldwell et al. 2006). But the roots of this shift are almost always analyzed from a social, political, or economic perspective, with the assumption that individual decisions about reproduction are what determine the fertility rate. This is despite the fact that, for any other species, the usual assumption is that ecological factors determine fertility rates.

The Moses and Brown article puts humans on the exact same scaling curves as any other species, subject to the same kind of cold, hard numerical analysis of population behavior that would apply to any other set of ecological data: "In our analysis, parents have as many children as they can afford to provision with the energetic resources expected in their society. As the cost and time to obtain these resources increase in more industrialized nations, the number of children parents can support decreases" (Moses and Brown 2003: 299).

Moses and Brown's study thereby flips the usual assumption about reproductive decisions. Most of the social and economic–based demographic studies assume that greater wealth brings about the luxury or ability to regulate fertility and to therefore decide to have fewer children. But this study suggests that greater wealth so constrains the energy-extended thirty thousand–kilogram virtual individuals that they cannot reproduce at a higher rate. The implications for the Global South–Global North binary are weighty. We need to think about not only how economic poverty might keep Global South populations from acquiring sufficient resources but also how wealth might constrain Global North populations from entering particular ecological niches or from changing certain behaviors that affect their environmental relationships. Under this rubric, in order to reduce overall energy use, we need a program of conservation and development work, based on a reverse South–North transfer of technology and cultural values, to be undertaken in North American suburbs.

Rather than working toward reducing fertility rates among economically poor Southern human populations, conservation goals and the recognition of the sacred might be better served if Global North populations' energy consumption is reduced to the point at which families can provision many more children. The actual power use of those additional children will be far less than the thirty thousand–kilogram individuals that a typical energy-extended North American family is currently raising.

In China, the legislated reduction of fertility rates (the one-child-per-couple law, which was instituted in 1979 and began to be phased out in 2015) has quickly brought population growth down; this is generally cited as one of the great goals of sustainable development and environmental conservation efforts. However, perfectly reflecting Moses and Brown's scaling relationships, this reduction in Chinese fertility has come with the expected scaled increase in energy use (U.S. Energy Information Administration 2014). In fact, some argue that acquired economic wealth, as much as the one child law, has been the key factor in reducing China's fertility rate. So, the question for environmental sustainability and conservation proponents is which goal is more important: reducing population growth rates or reducing energy consumption rates? It is probably impossible to have both.

Choosing between these priorities results in almost diametrically opposed sustainable development initiatives: the "reducing population growth" goal justifies typical programs that provide birth control and family planning for economically poor Global South populations, whereas the "reduce energy consumption" goal suggests that the correct course of action would be reconvincing economically wealthy Global North populations to devote more of their available metabolic energy to raising larger numbers of children. This in turn should result in a reduction in extrametabolic energy consumption as the family—not work, entertainment, and shopping—becomes the focus of a fulfilling life. Behind what we often believe are these apparently cultural choices, then, may well be the same biological imperatives, some would say "laws" regarding the relationship between birth rate and energy use.

It remains to be seen whether humans, using the potential for consciously changing the way we behave in the face of a biological imperative, will demonstrate lower birth rates along with lower levels of energy use. Part of what we might call the preservation of indigenous peoples and sacred sites often involves a call for reverence, or at least valuation, of a low-energy lifestyle and even low-energy settlements. However, when living peoples are involved, can we ask them to forego the quest for higher energy use, almost always accompanied by lower birth rates and higher education rates? Or might indigenous peoples find a way to lower birth rates and lower energy use while preserving some cultural features?

Perhaps more important, however, is whether a focus on asking Northern "modern" populations to be more like those of the low-energy South, or expecting high birth rate populations in the South to reduce their fertility, will successfully reduce suffering. The question is posed by the hard numbers of a well-done ecological scaling analysis but best answered by a softer science approach in which we try to find common ground between the realities of suffering and fulfillment experienced by people at far ends of the energy use spectrum. And this is where our South–North binary may provide the kind of balance and the kind of cultural perspective that is invaluable. When we employ the words "indigenous" and think of certain sites as "sacred," we must take into consideration whether our presumptions about where that population lies on the energy resource curve influence how we label them.

Are we using these terms to try to give value, in a strictly nominal sense, without allowing the populations to create their own real power–economic value? For example, the case of Northern American Indian gambling may merit in-depth analysis in reference to these questions. Where a population lies on the resource use continuum, how the power use continuum affects rates of population growth, and where the cultural values of South versus North fit in this economic–ecological analysis all carry weighty implications for how we define the sacred, what it means to be indigenous, and what we mean by the conservation of indigenous and sacred sites. In the next section, we analyze these key terms for this volume from the perspective of the North–South power use binary.

The Indigenous, the Sacred, and Conservation in the Twenty-First Century: How the Binary Viewpoint Can Guide Us

As the forces of urbanization—"Northernization" and the great demographic shift from low to high power use, inevitably accompanied by the shift from high to low fertility rates—formulate the primary influences on human ecologies in the twenty-first century, what is the role for the South–North binary in framing this great transition, and at times ameliorating its potential deleterious effects? In particular, what light does our power use analysis shed on the meaning of "the indigenous and the sacred," and can the cultural frame of South–North help give a humane, respectful, and enlightened understanding of these concepts? In the following section, we explore several issues that help us to gain perspective of the question of how South–North cultural dynamics, as well as the accompanying biological reality of resource consumption and population growth, affect the way we define the terms "indigenous" and "sacred."

Are Indigenous Peoples Best Conceptualized As Low Power Use Populations?

Throughout the world, an intuitive understanding of "indigenous" suggests the people who were present in a region before others arrived bearing new high-energy use technologies. Nowhere is this intuitive understanding more obvious than in the Americas, where a continent fully populated by great civilizations was, and in some regions still is, conquered by those (typically European or people of European descent) bearing higher-power use extrametabolic technologies. Originally in the form of wind power (such as intercontinental ships), animal power (such as horses), and fuel wood power (such as forged steel), the Industrial Revolution quickly brought fossil fuel–based, extremely high power use technologies to bear on the conquest. These technologies included high-speed transport (railroads), mechanized farming (with the use of draft animals, forged plows, and tractors), and machine fabrication (as in the textile industry).

Indigenous peoples have typically and intuitively been considered those who predate these high-energy technologies, somehow a representation of what *was* in the times *before*. And they are often considered most authentically indigenous (at least from a Global North perspective) if they still maintain a very low energy use, low technology profile as hunter–gatherers or hand farmers. But is this power use analysis sufficient to understand what "indigenous" means in the Americas? As we see increased and mainstream interest in what has been called an "indigenous revival," to what extent can we understand this newfound interest as being based on increased power and the entry into a more globally North lifestyle among indigenous peoples? Can the South–North binary shed more light on these questions?

From a Global North Good Life perspective, the low-energy indigenous way of life is romanticized as an authentic way to live, and great efforts are now made to conserve this way of life, with preserves and museums, working heirloom farms, and large isolated nature reserves, all trying to isolate and/or highlight the salient aspects of the indigenous lifestyle. Among the most dedicated conservationists and environmentalists (representing a very well-developed Global North Good Life perspective), this way of life may even be viewed as the solution (or even "salvation," as the most strident environmentalists would phrase it) to our most pressing environmental problems. The low energy use lifestyle of indigenous peoples is admired as a noble and ancient stalwart (as in the stereotypical "noble savage"). In this view, modern peoples are seen as corrupted, having lost their ancestral purity, and are expected to sacrifice something, even if it is part of their social capital, toward conserving the scant evidences and artifices, or living pockets, of indigenous life still remaining.

The modus operandi, then, for a Global North approach to indigenous conservation is to create reserves, such as the famous reservations for North American indigenous tribes; or to include indigenous peoples within a larger sacred natural site designation, such as many National Parks in the Amazonian region; or to attempt to re-create indigenous lifestyles in a museum or living heirloom farm environment.

On the one hand, a good argument can be made that these are important undertakings that at least keep the essential elements of indigenous life before us, available to study and understand, and that they have prevented the entire destruction of a grand way of life that existed before the arrival of Europeans. Furthermore, the Global North mindset, although epitomized by the Good Life and pursuit of material goods, also includes, perhaps as a deliberate counterbalance to the Good Life, a strong social justice component. Unlike attitudes among many other conquering peoples around the world, this emphasis on social justice strives for respect and co-understanding with those who have been conquered, have less access to technology, and are at lower power-use levels.

On the other hand, it could be argued that the attempt to conserve indigenous low power use lifestyles only belittles the people who live them and that we gain little from trying to simply understand, or even emulate, their technologies from the past. Such conservation practices also fail to see that the most appropriate way to honor or respect those who have less power use at their disposal is to allow their own self-realization and destiny making, to choose for themselves which aspects of their indigeneity, or their ancestral cultural practices, they wish to carry forward into the modern world. From a Global South perspective, this way of understanding the "indigenous" has allowed some groups to enter and dominate the highest levels of power use, economically and politically.

A great example includes the Otavalo Indians in Ecuador, who have used their business acumen to assemble high levels of urbanized wealth while traveling the world and maintaining an artisanal presence, based on their handicrafts and musical abilities, in hundreds of the great cities of the world. Emblematic of the indigenous revival in the Global South is the 2005 Bolivian election of Evo Morales—the first self-identified indigenous president of an American country—and the important political power that indigenous parties now carry in many Latin American countries, especially the Andean countries. The central involvement of ideas about *el Buen Vivir* in the Global South indigenous revival indicates a desire to merge economic success with recognition of the importance of traditional cultural values.

In North America, indigenous revival has led to a takeover of management of many conservation and preservation sites and a self-generated pride in ancestral cultural practices, especially production of fine arts. These activities are seen as culturally important but also as a way to move toward increasing material and economic wealth. In addition to reclaiming the economic value inherent in the conservation of traditional cultural practices, adoption of the casino business has provided a new source of economic wealth. Many tribes carry on a discourse as to whether these activities, in particular the opening of a casino, are damaging to their traditional way of life.

The North–South binary, then, illustrates the idea of indigeneity—although it may be partly understood as where an ethnic group stands, historically and economically, on the power use continuum and also carries an important cultural component. The Global North worldview, with its focus on material wealth (and the Good Life), views indigenous people simultaneously as a relic from the past and as an important potential source of information (and thus power, albeit retrograde). The conservation, or maintenance, of indigenous lifestyles is then seen as an economic sacrifice that is undertaken for its potential social good, and even perhaps ultimate economic benefit based on the value of the obtained and preserved information.

The Global South worldview, on the other hand, sees indigenous people as part of a living, timely, natural–cultural, human-involved bioscape. With quality of life derived from *el Buen Vivir,* the indigenous way of life is not seen as something detrimental to higher-performance economic activity, but rather as a welcome addition to a complete picture of a fulfilled life, a picture that includes many noneconomic components that derive from the indigenous worldview. By understanding these two approaches to indigenous revival as a North-South binary, we are better able to see potential sources of conflict regarding indigenous issues in today's world and to seek inclusive solutions to those conflicts.

Are Sacred Sites Best Conceptualized As Low Power Use Regions?

Sacred sites are often associated, especially when culturally defined, with the current or past presence of indigenous people, therefore reflecting a low power use regime at the site. This is true for reserves, preserves, and parks as well as more urban, small town, and rural settings where sacred sites most often carry historical significance and are often managed with low levels of extrametabolic energy, reflecting their past use. Sacred natural sites that are fundamentally devoid of human presence obviously show zero levels of extrametabolic power use, since this is always human based.

As with ideas about indigeneity, whereas an initial understanding of sacred sites might come from a power use analysis, the Global South–Global North binary provides a richer picture of how sacredness is determined and what conservation of a sacred site might mean. For the Global North and its emphasis on economic well-being, sacred sites are most often understood as leisure-oriented tourist attractions, places people will visit when they are using their accumulated resources in a "nonproductive" fashion. The historical monuments, national and state parks, and working heirloom communities and farms are seen as quaint and informative entertainment undertakings and as a way to enrich our understanding of who we are and where we come from, albeit from a nonessential, and nonproductive point of view. (It is worth remembering, however, that the hard facts show tourism, which often is stimulated by sacred site visits, to be a very productive part of the overall economy, even though Good Life individuals do not always see it that way.)

From a Global South perspective, with its emphasis on including creative, family-based, and cultural elements as part of human quality of life, in the context of *el Buen Vivir,* sacred sites become incorporated into a daily assemblage of activities in which work is not clearly separated from leisure. Historic city centers might still be in use (the designation of the Quito-Ecuador Historical Center, a vibrant and crowded urban environment, as a World Heritage Site is a good example), and even rural or remote sites might be adapted with modern technologies. Boundaries around indigenous homelands, natural sanctuaries, and government-run parklands may not be well defined. The economic use of those lands may also be openly or discretely understood to be part of how a more rural and distributed population earns its living.

The South–North binary, again, permits a more robust appreciation of the conflicts that can arise around how sacred sites are defined and managed, with the hope that a more inclusive solution to those conflicts can be found.

Conclusion

As the indigenous revival movement spreads across the Americas, and the designation of a particularly important natural or cultural site as sacred becomes more inclusive, it is crucial to formulate a framework for viewing, understanding, and creating consensus about what it means to be indigenous, how a site is determined to be sacred, and what the value is of conserving such sites. Historically, the Global North has imposed an economic value, consumerist-based method for designating, creating, and managing sacred sites as preserved areas. The Global South, on the other hand, has more thoroughly integrated sacred sites into the daily rhythm of life in which preservation may be less successful, but the incorporation of noneconomic values into quality of life is profoundly appreciated.

Much of the difference between the Global South worldview (which we encapsulate in the notion of *el Buen Vivir*) and the Global North worldview (which we characterize with the idea of the Good Life) can be understood as part of the demographic transition from low to high power use that accompanies falling fertility rates. Global South populations tend toward the high-fertility, low energy use, historically earlier part of this transition, whereas Global North populations have transitioned into a realm where they consume extrametabolic energy (mostly in the form of fossil fuels) as if each individual weighed thirty thousand kilograms and have fertility rates below two, which precludes population replacement over the long run. This ongoing demographic transition will reach its zenith toward the middle of the twenty-first century, as fertility rates continue to drop in the Global South, finally resulting in an overall declining world human population size.

Although the demographic transition and its accompanying urbanization—with increasing value placed on leisure time and tourism to sacred sites as a definitive leisure activity—can explain some of the differences in how the indigenous and the sacred are understood by individuals in both the Global South and the Global North, embedded cultural viewpoints are also important. At any power use level, the Good Life calls for a distinct focus on material wealth, and the conservation of indigenous and sacred sites will come as a sacrifice to that ultimate life goal.

However, this sacrifice means that indigenous and sacred sites come to hold a very special place that leads to their successful preservation or conservation in situ with little change. By the same token, regardless of the power use level obtained by a Global South population, basing cultural roots on the notion of *el Buen Vivir* ensures that the indigenous and the sacred are adopted as part of everyday life; these provide essential cultural elements—such as creativity, family relations, and connectivity with the past—that are considered part of overall quality of life.

By offering this North–South binary perspective on the indigenous revival and the conservation of sacred sites, we hope to bring a new way of addressing and resolving many of the traditional conflicts that arise around these issues.

Esmeralda Guevara, Ph.D., is an instructor of environmental education, School of Gardens, Dade County, Miami, Florida. She is an expert in educational approaches for the personal development of children and in pedagogy of the Andean philosophy of good living.

Larry M. Frolich, Ph.D., is a professor at Miami Dade Community College, Miami, Florida, an editor of the *Journal of Sustainability Education,* and an expert in herpetology, conservation, and environmental communication.

Notes

1. Wikipedia, widely accepted as our best digital era encyclopedic compendium of important topics, creates such "disambiguation" entries when a particular topic title is ambiguous because of multiple references. For the case of "the Good Life," almost fifty Wikipedia entries, from film, literature, television, and music, contain the phrase in their title.

References

Ariza, Marina. 2003. "La Urbanización en México en el Último Cuarto del Siglo XX." Presented at the Latin American Urbanization in the Late Twentieth Century Project, August 2003. Montevideo, Uruguay.

Belk, Russell and Richard W. Pollay. 1985. "Images of Ourselves: The Good Life in Twentieth Century Advertising." *Journal of Consumer Research* 11, no. 4: 887–897.

Caldwell, John C., Bruce K. Caldwell, Pat Caldwell, Peter F. McDonald, and Thomas Schindlmayr. 2006. *Demographic Transition Theory.* Dordrecht: Springer.

Cardenas, Carla. 2010. "Un Nuevo Concepto Dentro Del Derecho Ambiental: Los Derechos de la Naturaleza" ["A New Concept in Environmental Law: The Rights of Nature"]. *Journal of Sustainability Education,* 9 May 2010. Available at: http://www.jsedimensions.org/wordpress/content/un-nuevo-concepto-dentro-del-derecho-ambiental-%E2%80%9Clos-derechos-de-la-naturaleza%E2%80%9D_2010_05/ (accessed 4 March 2012).

Centre for Bhutan Studies. 2012. "Gross National Happiness." Available at: http://www.grossnationalhappiness.com (accessed 8 April 2012).

Fath, Brian D., Sven E. Jørgensen, Bernard C. Patten, and Milan Straškraba. 2004. "Ecosystem Growth and Development." *Biosystems* 77, nos. 1–3: 213–228.

Gilbert, Alan. 1993. "Third World Cities: The Changing National Settlement System." *Urban Studies* 30: 721–740.

Global Footprint Network. 2012. "Do We Fit on Our Planet?" Available at: http://www.footprintnetwork.org/en/index.php/GFN (accessed 5 May 2012).

Gordon, Brandilyn, Fausto Sarmiento, Ricardo Russo, and Jeffrey Jones. 2010. "Sustainability Education in Practice: Appropriation of Rurality by the Globalized Migrants of Costa Rica." *Journal of Sustainability Education.* Available at: http://www.jsedimensions.org/wordpress/content/cultivating-sustainability-pedagogy-through-participatory-action-research-in-interior-alaska_2010_05/ (accessed 7 April 2014).

Gudynas, Eduardo and Alberto Acosta. 2012. "La Renovación de la Crítica al Desarrollo y el Buen Vivir como Alternativa" [Renewal of the Criticism of Development and El Buen Vivir as an Alternative]. *Journal of Sustainability Education.* Available at: http://www.jsedimensions.org/wordpress/content/la-renovacion-de-la-critica-al-desarrollo-y-el-buen-vivir-como-alternativa_2012_03/ (accessed 7 April 2014).

Holling, Crawford S. 2001. "Understanding the Complexity of Economic, Ecological, and Social Systems." *Ecosystems* 4, no. 5: 390–405.

Holling, Crawford S. 2004. "From Complex Regions to Complex Worlds." *Ecology and Society* 9, no. 1: 11.

International Institute for Sustainable Development. 2012. "Canadian Index of Well-Being." Available at: https://uwaterloo.ca/canadian-index-wellbeing/ (accessed 23 September 2013).

Kneebone, Elizabeth and Emily Garr. 2010. "The Suburbanization of Poverty: Trends in Metropolitan America, 2000 to 2008." *Brookings Institute Metropolitan Opportunity Series,* no. 4. Available at: http://www.brookings.edu/papers/2010/0120_poverty_kneebone.aspx (accessed 5 April 2012).

Lewis, Martin W. and Karen Wigen. 1997. *The Myth of Continents: A Critique of Metageography.* Berkeley: University of California Press.

Liedloff, Jean. 1977. *The Continuum Concept: In Search of Happiness Lost.* New York: DeCapo Press.

Max-Neef, Manfred. 1992. "Development and Human Needs." In *Real-life Economics: Understanding Wealth Creation,* edited by Paul Ekins and Manfred Max-Neef, 197–213. London: Routledge.

McConnell, Campbell R. and Stanley L. Brue. 2005. *Economics: Principles, Problems, and Policies.* 16th ed. New York: McGraw-Hill.

Moses, Melanie E. and James H. Brown. 2003. "Allometry of Human Fertility and Energy Use." *Ecology Letters* 6: 295–300.

New Economics Foundation. 2012. "Happy Planet Index." Available at: http://www.happyplanetindex.org (accessed 12 May 2013).

Rosling, Hans, producer. 2009. "Let My Dataset Change Your Mindset." Available at: http://www.ted.com/talks/lang/eng/hans_rosling_at_state.htm (accessed 5 June 2010).

Sahlins, Marshall. 1974. *Stone Age Economics.* New York: Routledge.

Sarmiento, Fausto O. 2003. "Protected Landscapes in the Andean Context: Worshiping the Sacred in Nature and Culture." In *The Full Value of Parks: From Economics to the Intangible,* edited by David Harmon and Allen Putney, 239–249. Lanham, MA: Rowman & Littlefield.

Sustainable Society Foundation. 2010. Available at: http://www.ssfindex.com (accessed 8 April 2012).

Thomas, Lewis. 1974. *The Lives of a Cell: Notes of a Biology Watcher.* New York: Viking Press.

United Nations Development Programme. 2011. "Composite Indices—HDI and Beyond." Available at: http://hdr.undp.org/en/statistics/understanding/indices (accessed 25 April 2012).

U.S. Energy Information Administration. 2014. "Annual Energy Outlook with Projections to 2040." Washington, DC: U.S. Department of Energy. Available at: http://www.eia.gov/outlooks/archive/aeo14/ (accessed 17 January 2017).

Varela, Olmedo J. 1998. "Modeling the Changing Urban Landscape of a Latin American City: Lessons from the Past and Considerations for the Future of Metropolitan Panama City." Presented at Latin American Studies Association Chicago Meetings, Chicago, Illinois, 24 September 1998.

Vermeij, Geerat J. 2004. *Nature: An Economic History.* Princeton, NJ: Princeton University Press.

Waters, Tony. 2007. *The Persistence of Subsistence Agriculture: Life Beneath the Level of the Marketplace.* Lanham, MD: Lexington Books.

Wesselink, Bart, Jan Bakkes, Aaron Best, Friedrich Hinterberger, and Patrick ten Brink. 2007. "Measurement beyond GDP." Available at: http://assets.wwfza.panda.org/downloads/beyond_gdp.pdf (accessed 7 April 2014).

Wikipedia. 2012. "The Good Life." Available at: http://en.wikipedia.org/wiki/The_Good_Life (accessed 4 October 2012).

World Bank. 2010. "Wealth Estimates." Available at: http://web.worldbank.org/wbsite/external/topics/environment/exteei/0,contentmdk:20487828~menupk:1187788~pagepk:148956~pipk:216618~thesitepk:408050,00.html (accessed 24 May 2012).

World Bank. 2013. "Adjusted Net Saving—A Proxy for Sustainability." Available at: http://www.worldbank.org/en/news/feature/2013/06/05/accurate-pulse-sustainability (accessed 7 April 2014).

Yale University. 2005. "Environmental Performance Measurement Project." Available at: http://epi.yale.edu/ (accessed 15 January 2015).

Yale University. 2012. "Environmental Performance Index." Available at: http://epi.yale.edu (accessed 24 January 2013).

Chapter 4

Sacred Mountains

Sources of Indigenous Revival and Sustenance

Edwin Bernbaum

Sacred mountains have a key role to play in helping to renew and sustain indigenous communities and cultures around the world. They are among the largest and most prominent of all sacred sites, giving them a special status and importance in many religious and cultural traditions. As the highest such sites, they form natural links between heaven and earth, the sacred and the profane. On a practical as well as a spiritual level, they are major sources of water on which billions of people depend for their existence. Failure to preserve their environments can lead to the demise of local communities and the traditions they uphold. The resulting flooding, pollution, and loss of water resources can also wreak havoc on the lives of those who live in larger population centers downstream from sacred mountains.

If people live on a sacred mountain, we can clearly see direct connections between the sacredness of the mountain and the health and integrity of the local communities and traditions. The Kogi, for example, draw strength and inspiration from beliefs and practices that have come from centuries of experience living on the Nevado de Santa Marta in northern Colombia. These beliefs and practices, based on the sanctity of the mountain, have inspired their priests to take on the role of elder brothers entrusted with the task of warning the rest of the world about the perils of global climate change (Ereira 1992). The Ayllu, or interconnected communities who live on Mount Kaata in Bolivia, owe their resilience to traditions that view the sacred mountain as a human body that holds them together (Bastien 1978).

Even if people do not live on a sacred mountain, it can still play a significant role in their communal lives, though a direct connection with the mountain may not be as immediately apparent. The Hopi of northern Arizona, for example, live on mesas more than one hundred kilometers from the San Francisco Peaks; however, what happens on the distant sacred mountain can influence the *katsinas,* ancestral deities on whom the Hopi depend to bring the rain-bearing clouds that enable them to grow crops in a harsh desert environment devoid of other sources of water.

Ski runs on the forested slopes of the San Francisco Peaks have defaced the mountain and interfered with the ability of the Hopi to perform ceremonies that, in their eyes, give them the title to their lands—something the American system of landownership and law has great difficulty understanding (Bernbaum 1997). Noting that his people derive a sense of spiritual well-being from prayers and songs in which they visualize their sacred mountain in a state of perfection, a Hopi spokesman has remarked, "If I am not able to achieve this kind of spiritual satisfaction because of that [the scarring of the San Francisco Peaks], I have been hurt, I have been damaged" (Dunklee n.d.: 75).

This chapter focuses on the distinctive ways in which sacred mountains have sustained indigenous cultures and the ways that they can promote indigenous revival. It examines, in particular, examples of how sacred mountains have held indigenous societies together in the past and what they suggest for the future. As the Hopi example illustrates, it is important to document how damaging the environment and downplaying or ignoring the sanctity of sacred mountains has undermined the vitality of indigenous communities and the integrity of their traditions. From such research, we can derive valuable lessons on what not to do in promoting indigenous revival and, conversely, what needs to be done to sustain and revitalize existing communities and traditions.

Mountains are sacred for a variety of reasons, depending on the various ways people view them. In my research on sacred mountains around the world, I have found ten views or themes particularly widespread: mountain as high place, deity or abode of a deity, place of power, center, symbol of identity, ancestor or abode of the dead, garden or paradise, temple or place of worship, source of water and other blessings, and place of revelation, transformation, or inspiration (Bernbaum 2006). Among these themes, those that pertain more to communal concerns—rather than focusing on individual aspirations—would seem to have the greatest potential for sustaining indigenous communities and promoting indigenous revival. They include the following: deity or abode of deity, power, center, identity, ancestor or abode of ancestors, and source. A survey of a few examples around the world will illustrate the roles these views or themes have played in the past and suggest how they might guide efforts to conserve the environment and help renew traditional cultures in the future.

A prime example of the key role of the theme of identity is the relationship of the interconnected communities of the Ayllu of Mount Kaata to the sacred mountain on which they live. They view Mount Kaata as a human body whose features they identify with the various features of their landscape, such as lakes, pastures, fields, and villages. They feed the mountain with animal sacrifices so that it will remain healthy and provide them with crops and livestock. As a villager explained to the anthropologist Joseph Bastien:

> The mountain is like us, and we're like it. The mountain has a head where alpaca hair and bunchgrass grow. The highland herders of Apacheta [the upper region] offer llama fetuses into the lakes, which are its eyes, and into a cave, which is its mouth, to feed the head. There you can see Tit Hill on the trunk of the body. Kaata [the main village, in the middle] is the heart and guts, where potatoes and oca grow beneath the earth. The great ritualists live there. They offer blood and fat to this body. If we don't feed the mountain, it won't feed us. Corn grows on the lower slopes of Niñokorin [the lower settlement], the legs of Mount Kaata (Bastien 1978: xix).

The identification with the mountain has been so strong that it has held the communities of Mount Kaata together despite efforts over the centuries by Spanish rulers and the succeeding Bolivian governments to break up the unity of their land and society and force them to abandon their rituals of feeding the mountain (Bastien 1978).

The Kogi of northern Colombia view the Nevado de Santa Marta as the heart of the world, the central place from which the human race originated. Living there at the very center, the *mamas*, their priests, feel they have the authority and responsibility to warn their younger brothers elsewhere about the dangers of what they are doing to the environment and the climate around the world. The view of the mountain where they live as the heart of the world has given the Kogi the strength to maintain their traditional way of life and preserve their community in the face of outside forces seeking to destroy it. It has also become a source of indigenous revival and growth, attracting attention and serving as a source of inspiration and a model of environmental and cultural conservation for people in the modern world (Ereira 1992; Mansourian 2005; Rodriquez-Navarro 2009).

For centuries the Quechua people who live around the base of Cotacachi in northern Ecuador have derived their sustenance and livelihoods from the sacred mountain's glaciers and snow cover, which feed the springs and streams that provide life-giving water. With the advent of global warming, the glaciers have disappeared and snow cover has become sporadic, leading to worries about the future of their water supply. Many of the older, more traditional people blame these alarming developments on bad human behavior, especially deforestation on the lower slopes of the mountain, which has angered Mama Cotacachi, their name for the sacred peak and the deity it embodies. The change in color of the summit, from white to black, makes them fear that Cotacachi is losing its sanctity and withdrawing its blessings.

The traditional healers who derive much of their power from the sacred mountain feel that its striking change of appearance has impaired their ability to heal. The view of Cotacachi as a sacred source of water and healing that has played a key role in sustaining local Quechua communities is threatened, leaving the people concerned and apprehensive about their future (Rhoades, Zapata, and Aragundy 2008).

The views of mountains as ancestors, symbols of identity, and places of power play particularly important roles in the tribal lives of the Maori people of New Zealand. Each Maori tribe, and even subtribe, has its sacred mountain, which it views as one of the legendary ancestors who came to New Zealand in canoes that shipwrecked on the coast. These ancestors wandered inland and froze into place as mountains that give the particular tribes descended from them their present identity. This is so important that when Maori from different places gather, they introduce themselves with a formula that starts with "My mountain is such and such" and goes on to list in order their river or lake and then their chief (Yoon 1986).

In 1887, colonists were coming into New Zealand and threatening to buy up land for grazing sheep on the lower slopes of Tongariro, the divine ancestor of the Ngati Tuwharetoa tribe. The paramount chief, Horonuku Te Heuheu Tukino IV, feared that if the sacred mountain of his tribe were cut up into pieces, it would lose its integrity and the mana, or power he and his people derived from it, would be lost. An advisor suggested that in order to keep Tongariro whole, he should donate it to the British crown to make into a national park.

He did so, and Tongariro became the first national park in New Zealand and the fourth national park in the world, established only twelve years after Yellowstone National Park in the United States (Lucas 1993; Tumu Te Heuheu 2006). The Maori were involved in its management from the beginning, and their role was further strengthened in 1993, when United Nations Educational, Scientific and Cultural Organization (UNESCO) inscribed Tongariro National Park as the first World Heritage Site in the new category of associative cultural landscapes (Rössler 2005; Tumu Te Heuheu 2006).[1] Today the Ngati Tuwharetoa are among the most prosperous and vital Maori tribes in New Zealand, and the current paramount chief, Tumu Te Heuheu, has played a major international role as chairman of the World Heritage Committee.

At this point I would like to focus in more detail on examples of sacred mountains as sources of indigenous revival that I have had the privilege of working on in projects with national parks and indigenous communities in the United States. Hawai'i Volcanoes National Park lies on Kilauea and Mauna Loa, two mountains sacred to Native Hawaiians as the abode and body of the volcano goddess Pele and her relatives. Recognizing the cultural importance of the area, the superintendent invited a committee of *kupuna,* or Native Hawaiian elders, to advise the park on matters of concern to Native Hawaiians. The Kupuna Committee expressed strong interest in replacing a painting of Pele by an Anglo in the main visitor center with a more traditional depiction of the goddess. Among other things, the existing work of art lacked any reference to Hawaiian traditions and made her look European with her blond hair on fire.

I was directing the Sacred Mountains Program of the Mountain Institute at the time, and when Hawai'i Volcanoes National Park managers asked us to work with them, we made it possible for the park to put out a call for submitting traditional paintings of Pele. A competition was held in which the Kupuna Committee would select the winning entry to replace the painting in the visitor center. The two main newspapers in Honolulu published articles about the project on their front pages (Wilson 2003), and in no time art stores on the Big Island of Hawai'i were sold out of supplies. The park was overwhelmed by what they described as "a tsunami of art." They had been expecting a dozen or so paintings and received more than 140 submissions, forcing park staff to work twelve-hour days processing them, a task many of the employees, a number of them Native Hawaiian, found the most meaningful work they had done in their Park Service careers.

The Kupuna Committee chose a painting by a local artist depicting Pele according to tradition as a Polynesian woman with a calm, compassionate expression, holding an egg symbolizing her creative, rather than merely destructive, powers, and stirring lava around her with a ceremonial staff imbued with mythic references. The Volcano Art Center next to the visitor center had planned to display the remaining paintings but had room for only fourteen, so the main hotel and the geology museum joined the art center in hosting a month-long exhibit of sixty-seven of the submissions titled "Visions of Pele." The competition and exhibit generated a great deal of excitement and energized the Native Hawaiian community living in the vicinity of the park, infusing them with pride in their traditions (Spoon 2007).

The Sacred Mountains Program raised funds for the second and more expensive part of the project, the commissioning of a large outdoor sculpture that would portray the Hawaiian concept of *wahi kapu,* or sacred place, as it relates to the volcanoes Mauna Loa and Kilauea. Eighteen sculptors submitted proposals with models and diagrams, but the Kupuna Committee felt that none of them met their criteria and turned them all down. This turned out to have positive consequences for indigenous revival since the park superintendent backed the elders in their decision, thereby enhancing their standing and authority in the Native Hawaiian community.

The Kupuna Committee and the park put out a second call for proposals that included the following description of a dream one of the elders had had of how she envisioned the sculpture:

> Lava is flowing from Mauna Loa like a river. The upper part of a woman's body is visible in the lava flow—it's Pele riding down the flow, her eyes staring in anticipation, looking in the direction she's going to go. The body of Pele is not the whole body or like we think of a body. It's the upper torso only. Her hair is filling in behind her, also riding the flow, and she's looking out at the ocean. The lava flow, the image of the woman, is the volcano goddess who has come to show us, the people, her power. (NPS 2007)

The call asked sculptors to take the dream into consideration, showing the vitality of Native Hawaiian tradition as a living, evolving part of contemporary life.

This time the Kupuna Committee selected a winning proposal, and after two years of work, the sculpture was unveiled in a ceremony attended by a large crowd that heard representatives of the park, along with the sculptor, speak about the importance of instilling a sense of awe, reverence, and respect for the mountains and other natural features of the park. The seven-ton sculpture of volcanic rock carved in the shape of a mountain shows the face of Pele delicately etched on one side and a trickle of red representing lava flowing down the other side. Four panels of native koa wood around the base illustrate stories from the life of the goddess, past and future. The Kupuna Committee blessed the work with the name Ulamau Pohaku Pele, "Forever Growing Rock of Pele," and the park placed a wayside sign next to it explaining the importance of the concept of *wahi kapu* for Native Hawaiians (Wilson 2007).

The elders wanted to let visitors to the park know that they were entering a special place sacred to Native Hawaiians so that they would not treat it disrespectfully, as a mere recreation area. At a meeting of interpretive staff trying to figure out how to convey this message through signs outside the entrance, I suggested that they add a prelude about the special importance of Hawai'i volcanoes to the existing radio program that almost everyone driving into the park listened to for information on what to do and where to go. The interpreter in charge of the radio program was Native Hawaiian, and he composed in his own words the following introduction—preceded by the music of a traditional Hawaiian nose flute—which blends together in a particularly sensitive way the spiritual and physical characteristics of the park:

> Aloha and welcome to Hawai'i Volcanoes National Park. You may notice a change in the plant and animal life, climate, or maybe the way you feel as you enter the park. Don't be surprised; this is a common occurrence. For centuries people have felt the power and uniqueness of this place. Hawaiians call it a *wahi kapu* or sacred area. You are in the domain of Pele, the volcano goddess. She is embodied in everything volcanic that you see here. This is also home to a forest full of species that are found nowhere else on earth and two of the world's most active volcanoes. Hawai'i Volcanoes National Park is now a World Heritage Site, a modern term for a *wahi kapu,* recognizing its importance to all of us.

The introduction to the radio program provides a striking model of a way parks and protected areas can strengthen indigenous traditions and promote support for conservation of sacred sites at a cost of almost nothing.

An even more widely applicable model comes out of a collaboration that the Sacred Mountains Program initiated between Great Smoky Mountains National Park and the Eastern Band of the Cherokee (see Chapter 8). An easy 2.4-kilometer trail runs along the Oconaluftee River from a popular visitor center in the park to the Qualla Boundary, the ancestral lands of the Eastern Band, on the outskirts of the town of Cherokee in North Carolina. In addition to park visitors, many Cherokees walk this trail for exercise with their children. Coming out of a meeting in 2001 hosted by the Cherokee-operated Museum of the Cherokee Indian, the collaborative project led to the design and production of a series of wayside interpretive signs. These signs link features of nature along the Oconaluftee River Trail to Cherokee stories and concepts central to traditional Cherokee culture. Each sign is written in both English and Cherokee and is illustrated with a painting by a contemporary Cherokee artist. The stories and concepts are told in the words of Cherokee storytellers and elders in order to convey a sense of authenticity and immediacy.

Representatives of the Eastern Band chose the subjects and stories of the waysides in order to present what the Cherokees wanted park visitors to know about their traditions. An introductory panel at the start of the trail near the Oconaluftee Visitor Center explains the purpose of the signs and situates them in the context of the importance of sacred mountains in cultures around the world:

> As you walk the trail, you will encounter exhibits that contain Cherokee artwork, traditions, and quotations about the Cherokees' spiritual relationship with this place. These ancient mountains have long been home to the Cherokees, who honor the mountaintops as places to seek visions and receive direction from the Creator.
>
> People worldwide hold mountains to be sacred. Like the Cherokees, they connect their cultures' highest and most central beliefs to these dramatic landscape features.

Keep your spirits on the mountaintops.
Cherokee Shaman instruction

I will lift up mine eyes unto the hills, from whence cometh my help.
Psalm 121:1, *The Bible,* King James Version

As the dew is dried up by the morning sun, so are the sins of humankind by the sight of the Himalayas.
Hindu proverb

The succeeding waysides link natural features that are visible from each spot with corresponding Cherokee stories and quotations. For example, just beyond the visitor center one usually sees buzzards circling in the sky. The wayside points out the birds and tells the creation story of how, when the earth was soft and muddy, Buzzard flew out to dry it with his wings; where the wingtips went down he created valleys, and where they went up he made mountains. A sign on the banks of the Oconaluftee River with a Cherokee painting showing a face in the water includes the following text:

> Gunahita Yvwi—that means a long man. The river is the Long Man, with its head in the mountains and its feet in the sea. And its body grows as it goes along.
>
> The river was highly respected because it saves all life. Because if we didn't have water, everything would die—plants, animals, people, all things would be gone.
>
> And the Long Man was called upon for strength, for cleansing, for washing away sadness, for ailments. The water was used in so many ways. They had a lot of formulas and a lot of prayers that went over it. (Jerry Wolfe, Cherokee Elder)

Other waysides highlight the spiritual importance of water, trees, and a particular mountain for the Cherokee people. The introductory panel for those entering the other end of the trail near the town of Cherokee has the same explanation of purpose but adds a quote pointing out Cherokee reasons for protecting their mountains as sacred places:

> The Great Smoky Mountains are a sanctuary for the Cherokee people. We have always believed the mountains and streams provide all that we need for survival. We hold these mountains sacred, believing that the Cherokee were chosen to take care of the mountains as the mountains take care of us. (Jerry Wolfe, Cherokee Elder)

The signs were installed and inaugurated in 2006 with a ceremony that included traditional dances by Cherokee warriors and speeches by the chief of the Eastern Band, the superintendent of Great Smoky Mountains National Park, and other dignitaries (*Daily Times* 2006).

Since they walk the trail regularly with their children, the Cherokees see the waysides as a highly visible way of passing on their traditions to the younger generation and reinforcing the revival of the Cherokee language in their schools. The project also addresses a key concern of the Eastern Band: the need to promote exercise as a way of dealing with diabetes and other health problems arising from an epidemic of obesity. The Cherokees take pride in having developed a model that other tribes and indigenous communities can easily adapt to link their own stories and ideas to features of the natural environment that they hold sacred and that are integral to efforts to sustain their lives and revitalize their traditions (Bernbaum 2006).

The projects at the Hawai'i Volcanoes and Great Smoky Mountains national parks illustrate the key role that artwork and stories can play in sustaining traditional communities and promoting indigenous revival. This role brings up a basic point about values and knowledge that should be considered by organizations like the International Union for Conservation of Nature (IUCN) Specialist Group on Cultural and Spiritual Values of Protected Areas (CSVPA) that are working to highlight the importance of sacred natural sites and the need to take spiritual and cultural views into consideration in programs of environmental conservation. The following discussion of this point is meant to be exploratory rather than definitive.

Art and story, especially those of traditional cultures, do more than express values and evoke emotional responses to nature. They can change the way people see the world and reveal important aspects of reality that science misses, as science operates primarily through objective observations, repeatable experiments, theoretical concepts, and simplified models. In particular, works of art and stories can get below the superficial and give us a fuller, richer, and deeper experience of what is actually there in their concrete uniqueness and immediacy; we become acutely aware of features of nature and our relationship to them that we have overlooked or taken for granted. This deeper experience of reality provides the basis for knowledge of the natural world and lies at the heart of what connects people to nature, arouses emotions, creates and affirms values, and motivates conservation.

If organizations such as the CSVPA limit themselves to focusing primarily on the values of sacred sites and protected areas, they run the risk of inadvertently undermining their efforts to be taken seriously by scientists and protected area managers, on the one hand, and by the leaders of traditional communities and traditions, on the other. An exclusive focus on values plays into the tendency of scientists and managers to dismiss such approaches to environmental conservation as folkloristic icing on the cake, having to do primarily with how people feel about nature, rather than having anything to do with what is actually there. Science places great value on knowledge that is objective and free of values. If traditional views of nature are all about values, then they are merely subjective and have little use as valid sources of knowledge about the world and what needs to be done to protect the environment. They can be easily relegated to entertaining stories, picturesque ceremonies, and edifying works of art that provide a colorful backdrop to the real work of environmental conservation.

A focus on values can also inadvertently undermine efforts to gain the support of elders and other traditional leaders. Failing to acknowledge the reality that their views of nature and sacred sites have for them and their people can alienate them by making them feel they are not being properly respected. These views are not just a matter of values reflecting how they feel about things, but rather valid sources of knowledge and experience about the world with profound implications for the ways they live and treat the environment. At the very least, their traditions include practical knowledge about medicinal plants, climate patterns, animal life, and other matters that modern science finds extremely useful. At a deeper level, as the previous discussion of art and story suggests, indigenous views open us to direct knowledge and an experience of nature from which science tends to keep us removed, due to its emphasis on acquiring objective knowledge by separating the subject or observer from the object of observation.

The distinction is comparable to two kinds of knowing distinguished in the two verbs "to know" in Spanish: *saber* and *conocer* (the same distinction holds in French with *savoir* and *connaître*). The first kind, exemplified in *saber* and corresponding to scientific knowledge, is knowing about something or someone through description and explanation. The second kind, exemplified in *conocer* and corresponding to artistic and indigenous knowledge, is knowing something or someone experientially, as in the biblical sense of a man knowing a woman in sexual union. We can talk about knowing things about a person versus knowing that person personally: *Yo sé que él es Norte Americano, pero no lo conozco,* "I know that he is North American, but I don't know him personally." The same holds true of knowing a place, such as a sacred site, versus knowing about it. Each kind of knowing has its uses. For the fullest and richest possible knowledge of nature, we need both kinds of knowing—scientific on the one hand and traditional or artistic on the other.

The second kind of knowledge, expressed in the verb *conocer,* is important for conservation since it establishes an intimate connection with nature that motivates people to care for and protect the environment. Works of art and traditional views of natural sacred sites such as mountains help to overcome the subject–object dichotomy that separates us from nature and rationalizes environmental destruction and desecration in today's predominantly economic world. This is not a matter of being merely subjective, but rather one of evoking subjective experiences of an objective reality that reveals aspects of what is actually there that are not accessible to a purely objective approach to knowledge. We can see hints of the approach in physics as well with the Heisenberg uncertainty principle, which says that the observer has to be taken into account in observations at the quantum level.

As the examples in this chapter have shown, many traditional views of nature, especially of sacred sites such as mountains, imbue natural features with human personality, seeing in them the presence and work of spirits, deities, and ancestors. Anthropomorphic views of this sort help people to connect with nature in a particularly intimate way since people relate most easily to other people. This way of seeing the world can also help make people aware that they are part of nature, not just disembodied minds observing it from the outside.

At the very least, it highlights the ability of the environment to respond to what we do to it, just as another person might. Scientists tend to dismiss anthropomorphism as a primitive and erroneous view of the natural world that modern society has outgrown. This creates a resistance to acknowledging the validity of traditional sources of knowledge that needs to be overcome. But science itself is anthropomorphic in the sense that its experiments and theories are based on models created by and comprehensible to human beings. As Wendy Doniger, a leading scholar of comparative religion, put it in a class I sat in on as a graduate student, science projects onto the universe the model of a human legal system, seeing it as governed by laws in a way similar to the way societies are governed.

A fundamental key to success in efforts to promote indigenous revival and sacred sites conservation is the need to respect and take seriously traditional views of nature and the world. Toward this end, in addition to talking about values, I would suggest that organizations engaged in this work explore ways of elucidating and communicating the cultural and spiritual dimensions of experience and expression. They could also show how they can be valid and necessary sources of knowledge that complement scientific approaches to the protection of the environment in places that have special significance for indigenous peoples. Elders and other traditional leaders should play a leading role in these efforts and in developing and implementing programs that come out of them, particularly those that affect their sacred sites and traditions. To assist them in their work and to help communicate its importance to the wider public, the knowledge and skills of poets, artists, musicians, and scholars of the humanities would be invaluable in complementing the expertise of social and natural scientists.

Edwin Bernbaum, Ph.D., is co-chair of the International Union for Conservation of Nature Specialist Group in Cultural and Spiritual Values of Protected Areas. He is a senior fellow of The Mountain Institute, Washington, D.C. He is an expert on comparative religion and mythology with a focus on the relation of culture to the environment and the author of the award-winning book *Sacred Mountains of the World.*

Notes

1. Associative cultural landscapes are singled out for the cultural importance of their natural features, in this case a sacred mountain.

References

Bastien, Joseph W. 1978. *Mountain of the Condor: Metaphor and Ritual in an Andean Ayllu.* St. Paul: West.

Bernbaum, Edwin. 1997. *Sacred Mountains of the World.* Berkeley: University of California Press.

———. 2006. "Sacred Mountains: Themes and Teachings." *Mountain Research and Development* 26, no. 1: 304–309.

"Cherokee Exhibit in Park Dedicated—Tribal Leaders Give Blessing to Displays." 2006. *Daily Times,* 22 April, Stories.

Dunklee, John P. n.d. "Man–Land Relationships on the San Francisco Peaks." Museum of Northern Arizona Technical Series [unpublished]. Flagstaff, AZ.

Ereira, Alan. 1992. *The Elder Brothers.* New York: Knopf Doubleday.

Lucas, P.H.C. Bing. 1993. "History and Rationale for Mountain Parks as Exemplified by Four Mountain Areas of Aotearoa (New Zealand)." In *Parks, Peaks, and People,* edited by Lawrence S. Hamilton, Daniel P. Bauer, and Henry F. Takeuchi, 24–28. Honolulu: East-West Center, Program on Environment.

Mansourian, Stephanie. 2005. "Sierra Nevada de Santa Marta, Colombia." In *Beyond Belief: Linking Faiths and Protected Areas to Support Biodiversity Conservation,* edited by Nigel Dudley, Liza Higgins-Zogib, and Stephanie Mansourian, 114–116. Gland and Manchester: World Wide Fund for Nature; Equilibrium and the Alliance of Religions and Conservation.

NPS [National Parks Service]. 2007. "Hawai'i Volcanoes News Release: Community Invited to Join in Unveiling of Wahi Kapu Sculpture." Available at: http://home1.nps.gov/havo/parknews/loader.cfm?csModule=security/getfile&PageID=150203 (accessed 6 June 2014).

Rhoades, Robert E., Xavier Zapata, and Juan Aragundy. 2008. "History, Local Perceptions, and Social Impacts of Climate Change and Glacier Retreat in the Ecuadorian Andes." In *Darkening Peaks: Glacier Retreat, Science, and Society,* edited by Ben Orlove, Ellen Wiegandt, and Brian H. Luckman, 216–225. Berkeley: University of California Press.

Rodriquez-Navarro, Guillermo E. 2009. Personal communication.

Rössler, Metchild. 2006. "World Heritage: Linking Nature and Culture." In *Conserving Cultural and Biological Diversity: The Role of Sacred Natural Sites and Cultural Landscapes,* edited by Thomas Schaaf and Cathy Lee, 15–16. Paris: UNESCO Division of Ecological and Earth Sciences.

Spoon, Jeremy. 2007. "The 'Visions of Pele' Competition and Exhibit at Hawai'i Volcanoes National Park." *CRM: The Journal of Heritage Stewardship* 4, no. 1: 72–74.

Tumu Te Heuheu, Chief Tukino VIII. 2006. "Culture, Landscapes and the Principle of Guardianship." In *Conserving Cultural and Biological Diversity: The Role of Sacred Natural Sites and Cultural Landscapes,* edited by Thomas Schaaf and Cathy Lee, 224–226. Paris: UNESCO Division of Ecological and Earth Sciences.

Wilson, Christie. 2003. "Volcanoes Park Seeks New Portraits of Pele." *Honolulu Advertiser,* 27 May.

———. 2007. "Massive Pele Sculpture Unveiled." *Honolulu Advertiser,* 8 March.

Yoon, Hong-key. 1986. *Maori Mind, Maori Land: Essays on the Cultural Geography of the Maori People from an Outsider's Perspective.* New York: Peter Lang.

CHAPTER 5

Frozen Mummies and the Archaeology of High Mountains in the Construction of Andean Identity

Constanza Ceruti
Translated by Monica Barnes

Introduction

The Andes Mountains have played an important role in beliefs and ceremonial practices for local communities from pre-Hispanic times to the present. Andean communities attribute to mountains the power to impart fertility to flocks and harvests and to guard the mineral riches of their cores. The souls of the ancestors come to dwell on their summits, and people fear the volcanic eruptions, storms, and droughts with which the mountain spirits punish ritual omissions and transgressions (see Figure 5.1).

In the time of the Incas, high mountains were centers of religious pilgrimage. Their peaks were sites for the performance of rites, including incineration in ceremonial hearths, ritual burial of sumptuary objects, and human sacrifice. The bodies of children and young women who were buried there froze in the constantly low temperatures, and they have remained excellently preserved for more than half a millennium due to the dry, cold environment (Ceruti 2010b).

The Andean mountain sanctuaries are the highest archaeological sites in the world (Ceruti 1999). Traces of ceremonial activities are not found at higher altitudes on any other part of the globe, not even in the Himalayas, where sacred mountains are venerated by circumambulating their bases but not by ascending to their peaks. The Incas were the first to scale the highest summits of the Andean range. Today, these mountaintops are revered by local communities from prudent distances. However, the Incas faced the extreme rigors of the high-mountain environment, and they

Figure 5.1. A sacred volcano in northern Argentina (photo credit: Constanza Ceruti)

knew how to overcome the psychological barrier presented by the terror of the colossal peaks (Reinhard and Ceruti 2010).

Before the dawn of high-mountain archaeology, the greater part of our information on Inca religion and ritual was obtained from ethnohistorical sources. Effectively, material evidence of the Incas' ceremonial activities has rarely been preserved because, during the Spanish conquest and colonization, temples and ritual objects were systematically destroyed (see Reinhard and Ceruti 2006). However, neither the destructive fervor of the extirpators of idolatry nor the greed of the conquerors reached the high-altitude sanctuaries. Because of these factors, these sanctuaries constitute privileged sites in which unique material evidence of Inca sacrifices and offerings is preserved. The offerings of metalwork, ceramics, featherwork, textiles, and objects made of shell deposited in the high-altitude shrines are among the most varied and well-preserved collections of Inca artifacts that have survived (Ceruti 2003a).

The archaeological patrimony of the high mountains is susceptible to looting and intentional depredation by treasure hunters. It is also vulnerable to deterioration caused by recreational and extractive activities, in addition to destruction by global warming, glacial retreat, and natural

Figure 5.2. Andean pilgrims at the mountain shrine of Punta Corral (photo credit: Constanza Ceruti)

forces such as lightning storms. High-mountain archaeology enables us to protect evidence of considerable patrimonial and scientific importance, as demonstrated in the case of the three mummified children discovered at an altitude of more than 6,700 meters on Mount Llullaillaco in the Argentinean Andes (Reinhard and Ceruti 2000).

The frozen bodies of the young women and children offered by the Incas, because of their excellent preservation in the cold and dry conditions of the high mountains, provide unique opportunities for interdisciplinary studies integrating bioanthropology, paleopathology, and paleoradiology (Ceruti 2004b). These, and other specialties, prove the bioarchaeology coming from the high peaks to be an irreplaceable source of information on past human populations.

The Llullaillaco mummies, for example, were the object of radiological studies (Previgliano et al. 2003, 2005), odontological examinations (Arias, González Díez, and Ceruti 2002), and hair analysis to reconstruct the paleo diet (Wilson et al. 2007), as well as microbiological and pathological studies and ancient DNA studies, among others (Ceruti et al. 2008; Reinhard and Ceruti 2010). The potential of the frozen mummies as sources of knowledge will increase as various areas of investigation produce new scientific and technical advances. Their study allows a deeper comprehension of ritual in the time of the Incas, casting light on the identity, social status, ethnic affiliation, and state of health of those who were part of the Inca sacrifice ceremonies and offerings to the mountains (Ceruti 2005c, 2005b).

The importance of the mummies to the definition of Andean identity could remain unperceived or could remain difficult to understand, as it comes from foreign cultural contexts or belief systems. Familiarity with the dead is a characteristic sociocultural trait in various parts of Latin America, dating back to pre-Hispanic times and still extant today. In this context we must remember the active social and political role the mummies of the emperors played in Inca times; they were periodically paraded through the main plaza of Cuzco and maintained in palaces, enjoying the same attention that they received when they were alive.

We should also mention the spontaneous veneration that the naturally mummified bodies of certain infants receive today in the sierras of La Rioja, in western Argentina. These are believed to be religious intermediaries and are displayed for the devotion of the faithful, who pray to them as "little angels." The Inca mummies from the ice also fill a distinguished role in the construction of Andean identity and in the processes of "re-ethnicization" that are currently taking place in northern Argentina.

The international repercussions of the discovery and recovery of the mummies from the Llullaillaco volcano helped to lend visibility to the needs of the indigenous communities of Argentina, motivating governmental authorities to recognize their rights and inspiring society in general to become interested in their welfare (see Figure 5.3). Consequently, in recent years, numerous communities have coalesced and formed in several Argentinean provinces, in the context of a strong and sustained native revindication movement. In multicultural societies like that of Salta, whose social identity was traditionally anchored in the Hispanic arrival and the gaucho culture, the Andean cultural heritage has become substantially more highly valued. Since the discovery of the "children of Llullaillaco," importance has begun to be placed on the study of the Inca civilization both in Salta and in other parts of Argentina, as part of the basic content of the school curriculum. There has also been a noticeable and exponential increase in the number of university students studying anthropology or archaeology.

There has also been an increase in interest, on the part of the general public, in studying the Quechua (also spelled Kichwa, as in Chapter 10) language and pre-Columbian cultures. Likewise, there has been a notable augmentation in tourism in the region. This increases the possibility that young people born in rural areas can remain in their communities, instead of being forced to migrate to large cities in search of work. On the other hand, the mummies from the ice, and high-altitude archaeology in general, have contributed to a strong "mountain identity" in Salta, almost accidentally. The practice of the sport of mountain climbing has grown significantly, and the Andean heights have come to fill an emblematic role that can be appreciated in the current terms of "high-altitude tourism" and "highland wines," which are produced in the Calchaqui valleys. Likewise, the most recent works of many Andean painters and writers have been inspired by the figures of the "children of Llullaillaco," who occupy a unique and irreplaceable spot in the hearts and minds of the people of Salta.

Nevertheless, in spite of their indisputable scientific and patrimonial importance, the future of the Inca high-altitude sanctuaries and the high-altitude Andean mummies, in the majority of cases, remains uncertain. Diverse human and natural factors threaten their preservation to some extent.

Threats to the Preservation of the High-Altitude Archaeological Patrimony: A Diagnosis

To date, more than one hundred Andean mountains with sanctuaries on their peaks have been surveyed (see Ceruti 1998, 1999, 2001a, 2003b, 2004a, 2005a, 2008a, 2010a; Schobinger 1996). It is estimated that approximately 70 percent of the documented high-altitude archaeological sites show evidence of the impact of destructive processes (Ceruti 2004a). The preservation of the architectural and artefactual evidence of high-altitude sanctuaries has been found to be threatened due to human and natural factors.

Treasure hunters are known in the Andean world as *huaqueros.* As a consequence of their activities, excavations classified as "*huaquero* pits" are frequently encountered in the architecture of high-altitude sanctuaries (Ceruti 2007b). Consequently, it has been shown that as a result of *huaquero* activity in past decades, ever more damaging processes are affecting archaeological structures and bringing about the near-total destruction of mummified bodies and grave goods. This has been the case for the female mummy dynamited at a height of more than six thousand meters alongside an Inca platform near the snow-capped peak of Mt. Quehuar in northern Argentina (see Reinhard and Ceruti 2005; Ceruti 2001a). The mummy from the snow-capped Chuscha mountain, and the structure where it was originally buried suffered considerable *huaquero* damage during the early 1920s.

In addition to the intentional actions of treasure hunters is the unintended destruction of architectural patrimony that occurs through the establishment of mountain climbers' camps; these reuse archaeological sites with logistical functions found at the bases or on the slopes of the mountains. One also sees modifications to the architecture of the high-altitude sanctuaries made by the construction of commemorative rock piles on which Catholic crosses are planted or that are used to testify to sports ascents. On the mountains of intermediate height along borders, alteration of the peaks for the erection of geodetic markers has occurred. The collection of artifacts and fragments of wood as "souvenirs" also negatively affects the fragile archaeological record on the surface of peaks with high-altitude Inca sanctuaries.

The recent increase in mining activity and the practice of adventure tourism with the consequent movement of people into areas that had not been frequently visited generate a negative impact that puts the long-term conservation of high-altitude ceremonial sites in serious danger. There are numerous events and processes that threaten the preservation of mountain sanctuaries, including the opening of pits for geological exploration, the construction of mining camps in areas with archaeological occupation, slope erosion caused by four-wheel drive vehicles and motorcycles, the erection of stone markers on the summits, the extraction of minerals at high elevations, and the installation of antennae and relay stations on peaks near cities.

Difficulties in preserving the archaeological patrimony of high-altitude sites extend beyond the slopes of the mountains, reaching the display cases and collections of museums. Unfortunately, there are mummified bodies and grave goods from high-altitude sanctuaries that have been irremediably damaged by problematic display practices in past decades (see Ceruti 2001b; Schobinger 1995). In this sense, the extraordinary state of conservation of the frozen mummies from the Llullaillaco volcano constitute a happy exception to the rule and represent the fruit of the efforts and dedication of many people and institutions (Ceruti 2011b).

The scarcity of resources dedicated to archaeological research is also a factor that acts indirectly against the preservation of high-altitude sanctuaries. I have accomplished the greater part of the high-altitude archaeological survey without dedicated funding, through personal efforts and austerity.

Today, in Scandinavian and alpine countries, the archaeology of glaciers, also called "mountain archaeology," has considerable institutional support due to concern for the conservation of materials that become exposed due to glacial retreat and the progressive melting of high elevation snowfields. Climatic factors and global warming also affect the Andean mountains and their cultural heritage (Ceruti 2007a). Nevertheless, it is regrettable that in many parts of the Andes, scientific work is obstructed by a lack of institutional assistance and a scarcity of resources for archaeological research at high altitude, which is aggravated by growing restrictions and bureaucratic impediments. Academic interests and rivalries among universities also negatively affect the feasibility of developing joint or collaborative research.

Mitigation of Impact on High Mountains: Proposals

A census should be conducted of high-altitude archaeological sites on the basis of a systematic exploration of mountains, with the purpose of making architectural plans and taking photographs of remains identified on slopes and peaks. This should be combined with an evaluation of the incidence of postdepositional processes in each case. The task of scientific research with archaeological aims on high mountains should be adequately supported, both during exploration and in circumstances in which it is necessary to make rescue interventions.

The construction of shelters, the opening of roads and pits, the laying of high-tension lines, the erection of relay antennae, mining activities, and all other infrastructure work in the vicinity of high Andean mountains should be preceded by environmental impact studies designed to evaluate negative effects that could directly or indirectly affect the archaeological record. Research that plans to accomplish archaeological work on peaks of more than five thousand meters should be conducted only by professional archaeologists with proven track records and experience in high-altitude work. The difficult access to high-altitude sites should not be an excuse for the interference of recreational mountain climbers, amateur archaeologists, or professionals from other scientific disciplines, as this work requires great responsibility and must be undertaken and supervised directly by a professional high-altitude archaeologist.

In order to contribute to the long-term preservation of Inca high-altitude sanctuaries, it is necessary that the commercial use of these mountains by tour operators be regulated. The inclusion of mountains with Inca sanctuaries in adventure tours must be limited exclusively to those high-altitude archaeological sites that have already been thoroughly studied scientifically and for which a management plan has been drawn up with adequate guidelines for tourist use. The guidelines must correspond to the specific conditions of each mountain and thus guarantee the long-term preservation of its archaeological patrimony. Within the framework of a careful development of sustainable management of the high Andean mountains, responsible tourism may bring the local inhabitants opportunities to function as custodians and local guides.

Mountain guides who take clients on ascents of volcanoes and peaks with high-altitude sanctuaries should be specifically trained for the task. Unfortunately, in many cases the people who take on the role of guides in these mountains do not have any formal training. Even those who have received formal certification are often lacking the education necessary to appreciate high-altitude archaeological patrimony and to preserve it through preventative means.

Educating the whole community about its patrimony is a fundamental aspect of the effective preservation of the archaeological record in all of its manifestations, including the high-altitude sanctuaries. Nevertheless, it is regrettable that this subject is almost absent in the curriculum of the various levels of education. The inclusion of assignments, materials, courses, or special classes oriented toward the theme of cultural patrimony would substantially rectify the enormous deficiency in this field. It is also an urgent necessity that professors and teachers be trained, both in rural and urban contexts. Mass communication could reflect these initiatives, taking advantage of the increasing popularity of documentaries and scientific content. There could also be an effort to deal specifically with themes related to regional cultural patrimony in graphic and print media. Researchers in the field of anthropological sciences should consider the task of bringing their work to the general public to be a fundamental part of their academic activity.

High-Altitude Andean Sanctuaries As Cultural Patrimony: Suggestions for Management and Conservation

Local Andean communities continue to treat the mountains as objects of veneration through offerings of coca leaves, alcohol, food, and animal sacrifices. Even though the ceremonial platforms on the peaks of the highest mountains are no longer used in traditional worship as they were five hundred years ago under Inca rule, certain mountains of lesser elevation continue to be the destinations of Andean religious pilgrimages. This is the case with the pilgrimage of the Lord of the Snow Star (QoyllurR'iti) in southern Peru and with the processions to the sanctuaries of Punta Corral and the Abra de Punta Corral (Punta Corral Pass) in northern Argentina (see Figure 5.2).

In addition to the traditional Andean pilgrimages, in recent years there have also been planned ascents to peaks of considerable altitude, which are organized within the framework of the expanding process of re-ethnicization in northwestern Argentina (Ceruti 2013). This is the case with the annual ascent to the summit of Mt. Macón, in northern Argentina, that the

Figure 5.3. Inca offerings from Mount Llullaillaco (photo credit: Constanza Ceruti)

highland communities have organized for about the last ten years (Ceruti 2011a). These events help to generate a greater emotional commitment on the part of local residents to the mountains that surround them. At the same time, this creates new challenges for the preservation of the archaeological record of high-altitude sanctuaries.

The present ritual activity and the traditional beliefs concerning the Andean mountains take on a new meaning when high-altitude archaeology allows us to frame them in a perspective of more than half a millennium of cultural continuity. The frozen child mummies throw light on contemporary beliefs about the "miraculous little angels" in the cemeteries of northwestern Argentina, particularly taking into account the importance of children as religious mediators in the Andean world (Ceruti 2010c).

The presence of ceramic miniatures and Inca figurines offered at the high-altitude sanctuaries is better understood in the context of modern Andean ceremonies, such as the festival of Santa Anita in the region of Humahuaca, in which the devotees engage in ritual games and parodies using miniatures that embody the essence of their prayers. Oral tradition about sacred mountains and their archaeology add profound significance to the archaeological evidence discovered in high-altitude sanctuaries.

This is the case with the oral tale the "Daughter of the Inca King," which is told by the inhabitants of the remote valley of Cajón after the removal of the mummy of snow-capped Mt. Chuscha by treasure hunters in the first decades of the twentieth century.

The high-altitude Andean sanctuaries, usually placed above five thousand meters, cannot be made into site museums because of their inaccessibility. The impossibility of guaranteeing the effective custody of archaeological sites at extreme altitudes likewise requires, in some cases, the direct intervention of the archaeologist, in the form of a preventative rescue, to protect the artifacts that make up the high-altitude archaeological record. In contrast, some archaeological sites at the bases of high mountains, those that have been associated functionally with pre-Hispanic ascension routes to the summits, can be developed through the creation of site museums or other alternatives oriented toward their protection, preservation, and sustainable use.

Local and indigenous communities must be participants in the processes of conservation and development of the Andean cultural patrimony. Their active involvement in all relevant cases is fundamental. What follows are some suggestions and examples that by no means cover the entire range of possible courses of action. Strategies should take into account the expert's advice and the needs and preferences of the community in every case, and competent authorities should provide the necessary resources for their implementation. Researchers should offer advice in their areas of expertise.

In Andean rural areas, the establishment of museums in the open air is being promoted. In recent years numerous community museums have been opened in which the traditional knowledge of the local people and the results of archaeological and anthropological research can make tangible contributions to the construction and appreciation of local identity as well as improve the quality of the cultural experience of outside visitors. The site of archaeological remains preserved in situ, accessible via interpretative trails and walking tours guided by local residents, lends value to the archaeological record and rock art, while contributing to the preservation of contextual information.

This approach also decreases the damage caused by the nonsystematic collection of remains or the rubbing of pictographs, for example. The participation of indigenous community members is also fundamental to the production of leaflets and signage that reflect oral traditions as well as to the creation of contexts and labels in museum exhibitions and visitors centers. In this sense, the strategies and activities implemented some time ago in Australia for the preservation and development of the archaeological patrimony of the Northern Territory and of the central desert in collaboration with the local Aborigine communities are illustrative (Ceruti 2007b; Verschuuren et al. 2010).

Archaeological objects coming from high-altitude sanctuaries—in their characteristic forms, decorative motifs, and manufacturing techniques—can serve as inspiration for local artisans, helping to emphasize the traditional character of their ceramic, textile, and metallurgical production. In addition to the artisan's own experience, the results of archaeological research into textile and ceramic technology, the analysis of fibers, studies of the composition of ceramic pastes, metallurgical analysis, and other areas of inquiry can be useful (Bray et al. 2005).

The desire of visitors to participate in traditional Andean annual ritual celebrations such as Carnival and the feasts of patron saints is increasing. Unfortunately, the massive influx of tourists at these festivals can have a negative impact on the normal development of these events. One potential alternative is the reenactment of ancient ceremonies such as the Inca processions and rituals of dedication. The details of these can be reconstructed on the basis of the archaeological evidence from high-altitude sanctuaries and the accounts in historical sources (Figure 5.4). In this sense, archaeology and ethnohistory can contribute to the opening of an intercultural dialogue that permits the celebration of Andean cultural patrimony through reenactment.

Figure 5.4. Surveying an archaeological site above 6,000 meters (photo credit: Constanza Ceruti)

Concluding Remarks

The cultural patrimony of the high Andes includes the mountain sanctuaries of the Inca civilization, which constitute the world's highest ceremonial sites. The archaeological evidence on these peaks includes, in exceptional cases, offerings and ice mummies in extraordinary states of preservation. These have survived until the present in spite of the extirpation of idolatry campaigns and the cultural looting that accompanied the European conquest and colonization. Unfortunately, the archaeological record of the high mountains is increasingly threatened by climate change, ignorance, and intentional destruction.

This chapter offers some thoughts on the preservation and development of this patrimony based on my personal experience. I have more than twenty years of residence in the Andes and have engaged in professional activities in the fields of high-altitude archaeology and anthropology. It also includes suggestions for studies of environmental impact on heights above five thousand meters, for the regulation of touristic use of peaks with Inca sanctuaries, and for the sustainable development and joint management of Andean mountain patrimony. The collaboration of scientific researchers, local communities, tour operators, and government authorities is necessary to guarantee the preservation of mountain sanctuaries for future generations within the framework of responsibility for the legacy of those who preceded us.

Constanza Ceruti, Ph.D., is professor of archaeology at the Catholic University of Salta, National Council for Scientific Research in Argentina (CONICET). She is an expert on high-altitude archaeology and sacred mountains as well as a National Geographic Emerging Explorer and TED fellow.

References

Arias Aráoz, Facundo, Josefina González Díez, and Constanza Ceruti. 2002. "Estudios Odontológicos de las Momias del Llullaillaco." *Boletín de la Asociación Argentina de Odontología para Niños* 31, nos. 2/3: 3–10.

Bray, Tamara, Leah Minc, María Constanza Ceruti, José Antonio Chávez, Ruddy Perea, and Johan Reinhard. 2005. "A Compositional Analysis of Pottery Vessels Associated with the Inca Ritual of Capacocha." *Journal of Anthropological Archaeology* 24: 82–100.

Ceruti, María Constanza. 1998. "Prospecciones en Sitios de Alta Montaña en el Noroeste Andino Argentino: Informe Preliminar." *Tawantinsuyu* 5: 37–43.

———. 1999. *Cumbres Sagradas del Noroeste Argentino.* Buenos Aires: Editorial de la Universidad de Buenos Aires.

———. 2001a. "Aracar, Guanaquero, Rincón, Arizaro y Blanco: Prospección y Relevamiento de Santuarios de Altura en Volcanes de la Puna Occidental Salteña." *Relaciones de la Sociedad Argentina de Antropología* 26: 145–166.

———. 2001b. "La Capacocha del Nevado de Chañi: Una Aproximación Preliminar desde la Arqueología." *Chungara* 33, no. 2: 279–282.

———. 2003a. *Llullaillaco: Sacrificios y Ofrendas en un Santuario Inca de Alta Montaña.* Salta: Ediciones de la Universidad Católica de Salta.

———. 2003b. "Santuarios de altura en la Región de la Laguna Brava (Provincia de La Rioja, Noroeste Argentino): Informe de Prospección Preliminar." *Chungara* 35, no. 2: 233–252.

———. 2004a. *Arqueología de Alta Montaña en La Rioja.* Salta: Ediciones de la Universidad Católica de Salta.

———. 2004b. "Human Bodies as Objects of Dedication at Inca Mountain Shrines (North-western Argentina)." *World Archaeology* 36, no. 1: 103–122.

———. 2005a. "A la Sombra del Volcán Licancabur: Santuarios de Altura en los Cerros Toco, Juriques y Laguna Verde." *Xama* 15–18, nos. 2002–2005: 301–313.

———. 2005b. "Actores, Ritos y Destinatarios de las Ceremonias Incaicas de Capacocha: Una Visión desde la Arqueología y la Etnohistoria." *Xama* 15–18, nos. 2002–2005: 287–299.

———. 2005c. "Elegidos de los Dioses: Identidad y Status en las Víctimas Sacrificiales del Volcán Llullaillaco y de Otros Santuarios de Altura Inca." *Boletín de Arqueología* 7, no. 2003: 263–275.

———. 2007a. "Realm of the Ice-Cloaked Mountain Gods." *Explorers Journal* 85, no. 3: 36–37.

———. 2007b. "Sitios Rituales Incaicos en las Cimas de los Cerros Archibarca y Tebenquiche." *Boletín del Instituto San Felipe y Santiago de Estudios Históricos de Salta* 47: 67–78.

———. 2008. "Panorama de los Santuarios Inca de Alta Montaña en Argentina." *Revista Arqueología y Sociedad* 18: 211–228.

———. 2010a. "Arqueología en la Sierra Más Alta del Mundo: Santuarios Incaicos en el Cerro Negro Overo y Cumbre General Belgrano de los Nevados de Famatina (La Rioja, Argentina)." *Inka Llajta* 1: 225–243.

———. 2010b. *Embajadores del Pasado: los Niños del Llullaillaco y Otras Momias del Mundo.* Salta: Instituto de Investigaciones de Alta Montaña.

———. 2010c. "The Religious Role of Children in the Andes, Past and Present." *AmS-Skrifter* 23: 125–133.

———. 2011a. "Resacralizando el Espacio y Construyendo la Identidad: Ascenso Anual de la Comunidad de Tolar Grande al Santuario de Altura del Cerro Macón (Norte de Argentina)." *Inka Llajta* 2, no 2: 213–228.

———. 2011b. "Santuarios de Altura y Momias Incas en Salta." *Boletín de Estudios Históricos de Salta* 49: 232–247.

———. 2013. *Procesiones Andinas en Alta Montaña: Peregrinaje a Cerros Sagrados del Norte de Argentina y del Sur de Perú.* Salta: Editorial de la Universidad Católica de Salta.

Reinhard, Johan and Constanza Ceruti. 2000. *Investigaciones Arqueológicas en el Volcán Llullaillaco.* Salta: Editorial de la Universidad Católica de Salta.

Reinhard, Johan and Constanza Ceruti. 2005. "Rescue Archaeology of the Inca Mummy on Mount Quehuar, Argentina." *Journal of Biological Research* 80, no. 1: 303–307.

Reinhard, Johan and Constanza Ceruti. 2006. "Sacred Mountains, Ceremonial Sites and Human Sacrifice among the Incas." *Archeoastronomy* 19: 1–43.

Reinhard, Johan and Constanza Ceruti. 2010. *Inca Rituals and Sacred Mountains: A Study of the World's Highest Archaeological Sites.* Los Angeles: Cotsen Institute of Archaeology Press, University of California Los Angeles.

Previgliano, Carlos, Constanza Ceruti, Facundo Arias Aráoz, Josefina González Díez, and Johan Reinhard. 2005. "Radiología en Estudios Arqueológicos de Momias Incas." *Revista Argentina de Radiología* 69, no. 3: 199–210.

Previgliano, Carlos, Constanza Ceruti, Johan Reinhard, Facundo Arias Aráoz, and Josefina González Díez. 2003. "Radiologic Evaluation of the Llullaillaco Mummies." *AJR* 181: 1473–1479.

Schobinger, Juan. 1995. "Informe Sobre la Relocalización de un Hallazgo de Alta Montaña del Noroeste Argentino: La Llamada Momia de los Quilmes." *Comechingonia* 8: 47–67.

———. 1996. "Los Santuarios de Altura Incaicos y el Aconcagua: Aspectos Generales Interpretativos." *Relaciones de la Sociedad Argentina de Antropología* 24: 7–27.

Verschuuren, Bas, Robert Wild, Jeffrey McNeely, and Gonzalo Oviedo. 2010. *Sacred Natural Sites: Conserving Nature and Culture.* London: EarthScan.

Wilson, Andrew, Timothy Taylor, María Constanza Ceruti, José Antonio Chávez, Johan Reinhard, Vaughan Grimes, Wolfram Meier-Augenstein, Larry Cartmell, Ben Stern, Michael Richards, Michael Worobey, Ian Barnes, and Thomas Gilbert. 2007. "Stable Isotope and DNA Evidence for Ritual Sequences in Inca Child Sacrifice." *PNAS (Proceedings of the National Academy of Sciences of the USA)* 104, no. 42: 16456–16461.

CHAPTER 6

Sacred Sites and Changing Dimensions of Andean Indigenous Identities in Space and Time

Christoph Stadel

Introduction

The key themes of natural sacred sites, conservation, and cultural identities and revival of indigenous peoples are intimately interrelated, because in most traditional societies, indigenous spirituality and religious beliefs are deeply rooted in and closely linked to nature and its various manifestations. Sacred natural sites are locations where belief systems interact with nature, and various mythologies, theologies, philosophies, cosmologies, and ethics are related to natural phenomena. Sacred natural sites are among the earliest forms of protected areas, and in many cases, they have retained their function as well-guarded shields against blatant forms of environmental destruction, a heedless overexploitation of natural resources, and an erosion of local cultures. In the words of Verschuuren et al. (2010: 1), sacred natural sites "encompass the complex intangible and spiritual relationships between people and our originating web of life."

Prominent sacred mountain sites in North America include Mount Shasta in the Cascades, a holy site for different native tribes; the San Francisco Peaks, revered by the Hopi people; Mount Graham in Arizona, an Apache sacred site; and the Black Hills in South Dakota, a sacred site of the Lakotas. In Mexico, the volcanoes of Popocatepetl and Iztaccíhuatl are worshipped by indigenous people as principal seats of mountain gods. In the South American Andes, the prominent mountain peaks of Ausangate and Ampato in Peru, of Illimani and Mount Kaata in Bolivia, and of the Llullaillaco in Chile count among highly revered sacred mountains since pre-Columbian times.

Given the fact that native communities live in many cases in proximity to sacred sites, they should be recognized as their principal custodians. In the past, the traditional belief systems of indigenous peoples were often ignored or even suppressed, and their sacred sites tended to be forgotten, or they were "desanctified" by various reckless economic activities. The local people were often seen as a hindrance to rigid externally driven conservation programs, as well as to large projects aiming to exploit the hydrographic, agricultural, forestry, mining, and touristic potentials of mountainous areas. As a result, indigenous communities were often denied their role as responsible stewards of ecosystems and sacred sites. With the revival and greater empowerment of indigenous people, this deplorable situation has changed in many parts of the world.

Native people have made significant strides in raising their voices, in taking actions against external interferences and incursions, and in protecting their environment, identities and cultural heritage. The rights of indigenous peoples and recognition of their culture and empowerment have also been strengthened by new assertive articles formulated in national constitutions, as well as by declarations of international agencies. In the latter case, the United Nations Declaration on the Rights of Indigenous People (UNDRIP, Article 12) of 2007 (United Nations General Assembly 2007) was a major legal cornerstone for indigenous revival movements, which stated, "Indigenous people have the right to manifest, develop and teach their spiritual and religious traditions, customs and ceremonies; the right to maintain, protect and have access in privacy to their religious and cultural sites" (UNDRIP 2008 qtd. in Verschuuren et al. 2010: 1). Zoomers (2006: 1023) calls this new attitude "a more flexible and dynamic perception of indigeneity."

Thus, sacred sites, with their accompanying rituals and ceremonies, play a pivotal role in indigenous revival: either they are closely guarded secrets shielded from external interferences, or they are made accessible to visitors as testimonies of indigenous religious beliefs and culture. In the latter case, native communities have formed partnerships with external governmental and nongovernmental institutions such as the Nature Conservancy.

It has been widely recognized that effective collaboration and partnership between key knowledge holders and stakeholders, including indigenous communities, scientists, governmental and nongovernmental organizations, and some private initiatives, can be a successful path toward protecting the environment, toward respecting and sustaining the sacred sites, and toward culturally and economically supporting indigenous communities. Because the Andes constitute one of the largest and most diversified mountain realms, are the cradle of many successive great civilizations, and contain a large number of sacred sites and protected areas, the following reflections on indigenous identities are focused on this mountain realm.

Indigenous Andean Identity

Indigenous identities in the Andean realm vary in space and over time; their images, concepts, and dimensions are not static, but rather are constantly evolving (Ströbele-Gregor 1994; Wilson 2000; García 2005). In the literature, the cultures and identities of indigenous Andean people are frequently collectively referred to as *andinidad* or *lo andino,* and specific philosophies and cultural, economic, and societal concepts, norms, and practices have been identified as characteristic of *lo andino.* (Regalsky 1994; Estermann 1999; Gade 1999; Salman and Zoomers 2003; Stadel 2001, 2003; Zoomers 2005; Sarmiento 2013a, 2013b). But it must be noted that a broad spectrum of Andean identities and cultures exists and that indigenous people exhibit multiethnic, multilingual, and multicultural traits, including the fundamental differences between highland and lowland indigenous communities (Field 1994).

In speaking of indigenous cultures, there has been a tendency to focus on traditional cultural phenomena as idealized shields against the influences of "Westernization," modernization, capitalism, and the excesses of neoliberalism. But indigenous communities, throughout their long history of occupying the Andean environments and utilizing their resources, have always demonstrated their capacity to adapt to changing circumstances, test new opportunities, and adopt new resilience strategies. Certainly, the cultural term of *andinidad* is historically and spatially not restricted to the time and geographical realm of the Inca. Cánepa (2008) speaks of the "fluidity of ethnic identities" and Gelles (2002: 257) calls for a "more flexible and dynamic perception of indigeneity." Therefore, over time, the terms *andinidad* and "indigenous culture" have undergone considerable transformations in interpretation and assessment. Estermann (1999) and Zoomers (2005), among others, point out that the terms *andinidad* and *lo andino* are exogenous in origin and tend not to be used by local communities, except when pursuing joint political goals on wider regional scales.

Indigenous Andean people, especially those who have lived for centuries in rural mountainous areas of the Andes, share some traits in their philosophy, religion, symbols, myths, rites, societal norms, and livelihoods; in the words of Gade (1999: 36), "Many autochthonous elements, practices, strategies, and symbols, both material and non-material, make up the sum of *lo andino.*" The Andean culture is rooted in the ecological assets and economic opportunities of diverse but fragile mountain environments. The principal objective of farmers and pastoralists has been to utilize the natural potential of altitudinal belts, topographic niches, and local water, edaphic, and vegetation resources.

Because of this, a diversity of production processes, rather than specialization, is the declared strategy for minimizing natural and human risks (Regalsky and Hosse 2009). This complementarity (*complementaridad*)[1] of economic activities is, for instance, manifested in a mosaic of different crops even on small plots, in a mixture of irrigated and rain-fed agriculture, in various forms of field and crop rotations, and in a combination of field cultivation and pastoralism. On a wider scale, economic exchange systems exist between more distant mountain regions, between highlands and lowlands, and between the rural areas and the market centers (Phélinas 2002).

Changing Livelihoods, Changing Identities

From Diversity and Complementarity to Specialization

The complementary orientation of rural livelihoods is neither unique to the Andes nor static and solely traditional. In most mountainous regions of the world, families and communities have had to rely on a diversity of economic pursuits, and these activities have been subject to many changes. While many traits of a traditional complementarity in the Andes have persisted, others, at least in the more remote areas, have attenuated or even disappeared, while new forms of rural economic diversification have emerged.

Innovations in new forms of field cultivation and pastoral activities have been manifold and have resulted in different spatial and socioeconomic patterns within the rural landscape. Among the factors and forces altering the traditional picture of complementary agricultural pursuits are the changes in climatic parameters and in access to water resources, the latter in the form of either aggravating shortages or, conversely, new irrigation potentials. This has at times resulted in a new polarization between water-rich and water-poor regions and communities. In addition, improved rural roads and enhanced accessibility have encouraged a stronger market orientation toward agriculture and a weakening of subsistence-oriented economies.

Agricultural specialization is particularly noticeable in an increase of profitable cash crops such as onions, garlic, and new high-yielding varieties of potatoes; in an expansion of irrigated alfalfa fields to feed larger cattle herds; in growing milk production for the urban markets; in new demands for alpaca wool; and in a proliferation of poultry farms and greenhouses for export-oriented cut flowers. Furthermore, the impact of external actors in the form of economic corporations, government agencies, nongovernmental organizations, and formerly rural residents now living in cities or in foreign countries has contributed, mainly through investments in larger scale and specialized forms of agriculture, to a restructuring of rural economies. Gelles (2002: 258) notes, "Andean cultural production is dynamic and adaptive, providing orientations and identity for villagers as well as for migrants who transit different national and international frontiers."

Today, the complementarity of rural economies encompasses increasingly nonagricultural sectors. In many communities, agricultural incomes were traditionally supplemented with the production of textile products, pottery, or adobe bricks or with the wage labor of low-income workers (*peones*) and seasonal workers in plantations or in nonagricultural jobs. With the advent of tourism, in addition to expanding clienteles for locally produced handicrafts, some rural people have found employment in various sectors of the tourism industry. With enhanced mobility and improved transportation systems, many rural people are seeking temporary or permanent urban jobs, leading to substantial emigration, especially of younger people. This outmigration has contributed to a shortage of farmworkers in some areas; however, the remittances (*remesas*) have also generated additional funds for new rural investments. Zoomers (2006: 1032) points out, "Many activities are closely linked and mutually reinforcing ... some are linked in a complementary way ... while in other cases they are of a more substitutive nature."

From Reciprocity to Monetization

Another pillar of "Andeanity" (*andinidad*) is the economic and social concept of reciprocity. Reciprocity (*reciprocidad*) is based on the principle that work, services, goods, and gifts between people and regions are exchanged in equitable and just forms, guided by rigid norms and conventions. Again, this principle is not unique to the Andes, as it is also found in most pioneer societies. However, in the harsh environment of the Andean mountains it has long been an indispensable form of mutual support and a shield against penury and risk. The traditional form of economic reciprocity has been an exchange of goods and services (*trueque*). In the past, it was the preferred practice for exchanging agricultural products from different agricultural zones and for trading produce between farmers and pastoralists and between rural people and people in urban centers.

However, the increasing monetization of rural economies has greatly weakened this practice. In a sociocultural context, *reciprocidad* remains a cornerstone of the identity of village communities (*ayllús*). In various forms, members of the community are expected to participate in the collective construction or maintenance of community infrastructures (*minka*) or to assist other families on an individual basis in the building or repair of houses and to help them with their agricultural work (*ayni*). For community members (*comuneros*), this is a source of security, bonding, and strength based on the principles of mutual responsibility, solidarity, respect, and justice.

In her research in rural southern Bolivia, Zoomers (2006: 1033) found that "in all communities, people stressed the importance of social norms," that "all *comuneros* are supposed to participate actively in communal life," and that they "are obliged to attend when called by the authorities for communal task (*faenas*) such as clearing the irrigation channels, maintaining the roads and bridges, repairing the streets in the district, transporting materials and construction work." Although traditional support systems and economic forms of reciprocal relationships generally appear to have weakened in the Andes during recent decades, each *comunero* is still embedded in the strong sociocultural context of his or her community and is expected to make an appropriate and substantive contribution.

Failing to fulfill communal obligations is sanctioned with a loss of esteem and reputation, penalties, or even virtual expulsion from the community. Rist (2000: 310–311) poses the question of "whether today reciprocity should be considered as a marginal, traditional relic, or as a modern strategy to escape poverty and to attain development." He concludes that *reciprocidad,* at least from the perspective of the local actors, remains today not only as a vestige of cultural heritage but as a contemporary and important economic and social pillar for Andean communities.

Andean Cosmo-Vision: Nature-Based Spirituality

Lo andino manifests itself spiritually in a specific cosmic vision and order (hereafter referred to as "cosmo-vision") for which Estermann (1999: 162–163) introduces the term *achasophie,* or "the philosophy of *pacha,* the universe with its array of spatial and temporal dimensions." In this Andean cosmo-vision, the daily life of the farmers (*campesinos*) is greatly influenced by nature, the social norms and practices of the community (*pachaqamachana*), and the spiritual life of the people (*pachaqamaq*). For the indigenous people (*runa*), land (*pachamama*), rain and water (*para*), vegetation (*sacha*), and nature in general (*sallqa pacha*) are of vital importance.

For an Andean person, *pachamama* is an organic, living entity with which he or she has a close material and spiritual relationship. A *campesino* has deep respect for nature, the soil, water, animals, plants, and natural phenomena. Therefore, this attitude also has an ethical dimension: humans are the custodians of the earth and have an inherent responsibility to protect and sustain it. They have to "listen" to nature and its manifestations, and they have to maintain close contact with the earth by numerous ritual ceremonies. This explains the fact that Andean native identity is closely tied to an attachment to the soil, the land, and the local territory and its resources; a severance of these ties implies for individuals, families, and communities a great material and spiritual loss and a severe impairment of heritage and identity.

Andean culture, knowledge (*saber andino*), and wisdom are rooted in a collection of experiences, practices, norms, and livelihoods. These have been transmitted from past generations but have also been adapted to new challenges and opportunities and modified by external influences. Andean culture is highly heterogeneous and multifaceted, and includes many variations of experiences with different natural environments, historical–cultural experiences, specific cultural traditions, economic pursuits, and types of external influences. Also, *lo andino* is characterized as "dynamic and evolving, still adapting to frameworks of global environmental change" (Sarmiento 2013a: 316).

Andinidad and Sacred Sites: Mount Ausangate, Peru, and Mount Kaata, Bolivia

Andean mythology, religious beliefs, rituals, and ceremonies are intimately associated with sacred mountains. One of the most prominent sites in the Andes is Mount Ausangate (at an elevation of 6,384 meters), which is located in the Vilcanota range of Peru. This impressive mountain has been a major shrine since the Inca period (Bernbaum 1997: 180–183). Believed to be the seat of *Apu,* the great mountain god, and the dwelling of the spirits of the dead, it is a significant geographical expression of Andean indigenous mythology and identity. Every year, a week before Corpus Christi day, at a mountain site near Ausangate, indigenous communities celebrate *Quyllur Rit'í,* the "Star of Snow Festival." This celebration blends old Inca beliefs and practices with Catholic rituals and ceremonies (see also Borsdorf, Chapter 2 of this volume).

Thousands of pilgrims gather at a special mountain campsite and implore the gods for a rich harvest and for the health of their animals. The festival reaches its climax when the specially chosen *Ukurus* (bear people) climb up to sacred places on a glacier and bring back to the awaiting pilgrims and their home communities chunks of ice believed to have holy and medicinal properties.

The sacred Mount Kaata, located in a remote northeastern part of the Cordillera Real of Bolivia, is a major sacred mountain for the Qollahuaya people (Bastien 1978, 1985; Bernbaum 1997. They view the mountain as a human being that embodies the different aspects of the religious and social individual and the collective life of the indigenous communities. The Qollahuaya people conduct various sacrifices and celebrate other rituals and ceremonies here. But in contrast to the pilgrimage to the Ausangate, which now also attracts many tourists, little was known about this site before Bastien visited this area in the 1970s, and the sacredness of Mount Kaata has been largely reserved and preserved for the indigenous communities.

Poverty and Development in the Andes

Similar to the concept of *andinidad,* or *lo andino,* perceptions of "poverty" and "development" also diverge between outsiders (e.g., scientists, technicians, and government officials) and local populations (Zoomers 2005, 2006). Focusing on the reality of indigenous *comuneros* in the Bolivian sierra, Zoomers (2006) refers to the proindigenous reforms and asks the vital question of whether there is "an Andean way to escape poverty," in spite of the ethnic revival in Bolivia.

While the indigenous Aymara, Uros, and Quechua are characterized by special traditions and ways of life, generally referred to as *andinidad,* it is debatable whether this alone will help them to conquer the scourge of economic marginalization and poverty. The term *andinidad,* Zoomers states, is a heterogeneous and dynamically evolving concept, and the realities of Andean life have to be understood in the context of local geographies and interpretations. Contrary to general criteria and statistics on poverty, local perceptions and dynamics of well-being (*Buen Vivir*) and poverty have to be considered, and not merely in material terms.

Poverty, in the traditional Andean perception, is not a lack of possessions and wealth in monetary terms, but being ill, having no family, being a poor worker or drunkard, and having deficient social relations. Individual well-being is primarily derived from being respected in the community, and the *reputación* of a person is based on fulfilling one's promises and community obligations. Zoomers points out that the younger generation may speak of poverty, but the older generation seldom does.

Considerable variation in the perception of poverty, needs, and development priorities also exists among the different geographical contexts and cultural realms. In his research on environmental stress in a profile extending from the foot of the Chimborazo to the piedmont zone of the eastern Cordillera in Ecuador, Stadel (1989, 1991) examined the problems and needs of farmers in eight distinct environmental zones. Based on extensive interviews of low-wage laborers who worked on plots, he made a distinction between "actively perceived stressors," that is, those that were spontaneously mentioned by *campesinos,* and "passively perceived stressors," which rural people identified from a list of thirty-two potential stress factors.

In identifying the most severe problems and the most urgent needs, Stadel (1989, 1991) noted major differences (1) between the pastoral communities of the high-altitude *páramo* zones, the small landholders of the Sierra, and the pioneer land colonization farmers at the margins of the Amazon lowlands; (2) between the indigenous communities of the Chibuleos, Salasacas, or Shuar, and the mestizo populations; (3) between people living in proximity to the cities of Ambato, Baños, and Puyo or close to the highways and *campesinos* farming at peripheral and rather isolated locations; (4) between small landholders (*minifundistas*) and big landholders (*latifundistas*); and (5) between male and female respondents.

While the traditional Andean cultures have been frequently praised as fundamental human capital for the sustainable economic and social survival of indigenous communities, Zoomers (2006: 1037) asks whether Andean cultures are an asset that helps people to move out of poverty or whether they are a constraint on development. In her research on Bolivian highland farmers, Zoomers (2006: 1041) found that *comuneros* rely on a combination of natural and human assets. Often, indigenous people who have acquired some economic prosperity are tempted to use this as a springboard into the city. They also tend to develop new lifestyles, social practices, and norms, thus distancing themselves from the traditional concepts of *andinidad.*

Finally, Zoomers cautions that idealized or politicized interpretations of *andinidad,* in many instances, do not coincide with the reality of what she calls "lived Andeanism": "Rather than struggling to maintain their indigenous values, people are mostly concerned with how to survive, and with finding individual ways to escape from poverty ... People in the Andes have distinct traditions but there seems to be no 'distinct way' out of poverty" (2006: 1043).

In recent times, the plurality of Andean life appears to have increased, the contrast between the few people enjoying an upward economic and social mobility and those remaining at the same marginal level or experiencing downward mobility has widened, and the strategies to cope with a harsh environment and the precarious economic situation have diversified. This heterogeneity notwithstanding, highland communities encounter many major problems, including weather and climate risks, shortages of agricultural and pastoral lands, limited access to markets and low monetary returns for many agricultural products, an inadequate water supply, restricted employment and income opportunities, insufficient health care and education standards, and social problems and tensions.

Discussion: Shifting Demographics and the Evolution of the Concept of *Andinidad*

In view of these constraints and problems, it is debatable whether the traditional concept of *andinidad* and/or new approaches and strategies can adequately cope with all facets of this fragile natural and socioeconomic situation. Academics, politicians, and outside experts have often praised the value of traditional Andean economic and social principles and strategies, as well as the resilience and adaptive abilities of the local people (Argumedo and Stenner 2008; van Wijk 2010). In this vein, they emphasize the proven autochthonous use and management of natural resources; the endurance, knowledge, and skills of local people; the safety nets of their collective strength; the capital derived from agricultural and livestock products and other local goods; and the cultural heritage and values. However, because of the impact of both internal and external factors and forces, rural livelihoods have changed, and the concept of *andinidad* has to be modified and extended.

In particular, the various forms of massive migration flows and the different aspects of urbanization have resulted in a corollary of impacts, some of them detrimental and others supportive of rural livelihoods. Today, many families and communities rely on the incomes of people working periodically or permanently in the cities, in lowland plantations, in mines and oilfields, or overseas. In spite of their absence, many of them still consider themselves *comuneros*. They may continue to own some land in their home communities or make new investments in houses, land, and infrastructures, and they may attend and sometimes finance community celebrations while other family members fulfill community obligations. On the other hand, some villages have suffered from a depletion of manpower and young local leadership because of the outmigration of youth who leave the villages to pursue higher education in cities.

Urbanization has had similar consequences but has also generated a reverse flow of communication, investments, ideas, modernization, and capitalism from the cities to the countryside. As cities grow, and as some educated youth return to their villages, urban mindsets are making their way into the countryside, leading to modified Andean rural identities. For example, growing individualism is eroding the traditional collective solidarity of the *comunas*. More affluent villagers may also aspire to buy a house or apartment in a nearby city. Zoomers (2006) has reported that this is so in the Bolivian highlands, where rural people have acquired property in the city of Sucre. Gill (1993: 78) notes, "People hope to use their wealth as a springboard into the city, where they can escape the drudgery of rural life in the Andes."

In recent years, a more culturally appropriate and sensitive approach to development concepts and practices has been adopted in the Andes and elsewhere (Boelens and Gelles 2005; Andolina, Laurie, and Radcliffe 2009). Radcliffe and Laurie (2006: 231), in their plea for "taking culture seriously in development for Andean indigenous people" aim at developing "a postcolonial and poststructuralist account sensitive to the historically and geographically variable and contested nature of the connection of culture with development." They see culture as a "flexible and creative resource" that can be adapted over time and that can represent a "dynamic template for action," (ibid.: 245) combining tradition and modernity. In many development initiatives, based on a partnership approach between local communities and outside scientists, the heritage of the indigenous identities and values serves as an indispensable framework within which the modern concepts and strategies for technological innovation, modernization, and development are tested.

An example of one of the most successful and locally well-respected local initiatives in recent years is the Sustainable Agriculture and Natural Resource Management Programme in the Cotacachi region of northern Ecuador (Rhoades 2006). This long-term project focused on the sustainable management of natural resources and a strengthening of the small-scale agricultural potential of the region. In a holistic development approach, the native communities fully participated in all stages of the project and were asked for their input and feedback. The program was guided by the major goals and principles of environmental conservation, and it promoted the gentle and sustainable use of natural resources, safeguarding and strengthening cultural heritage, and exploring strategies for enhancing the rural livelihoods.

Conclusion

This discussion of the multifaceted images and dimensions of Andean indigenous identities has revealed that these are not uniform, static, and solely traditional in nature. With changing lifestyles and livelihoods, their cultural expressions, economic pursuits, and political aspirations have been modified by developments from within their societal framework and by external influences. These trends notwithstanding, indigenous forms of spirituality, societal norms, and agricultural methods have survived. Today, one can even discern a new and self-assured expression of native dignity and pride and a revival of indigenous values and ways of life. In an effort to escape poverty and powerlessness, indigenous communities have come to the conclusion that they do not have to give up their cultural identity and adapt and assimilate themselves to Western forms of "progress" but that they do have to seek flexible and dynamically evolving expressions of their indigeneity.

In a similar vein, indigenous sacred sites have retained their spiritual importance as identity markers (Sarmiento 2012) for native cultures. In the Andes and other parts of the world, mountains occupy a special position:

> Because of their power to awaken an overwhelming sense of the sacred, mountains embody and reflect the highest and most central values of religions and cultures throughout the world ... They are regarded traditionally as places of revelation, centres of the universe, sources of life, pathways to heaven, abodes of the dead, temples of the gods, expressions of ultimate reality in its myriad manifestations. (Bernbaum 1997 xxii–xxiii)

To indigenous communities, as encoded in many religious beliefs worldwide, mountain peaks, special rocks and landforms, old trees, groves, and forests, glaciers, springs, rivers, and lakes are considered sacred. Beyond these natural sites, constructed places—temples, monasteries, religious monuments, and shrines in particular—are also regarded as sacred; they are prime sites for special rituals and ceremonies and major pilgrimage destinations.

As traditional mountain cultures (and the indigenous religions originating there) are today at risk, the sanctity of these special natural and cultural sites may also be threatened. This may occur by monetization of mountain landscapes and ecosystems and by ruthless exploitation of the natural mountain resources. To counteract these developments, nongovernmental organizations, governments, and local organizations have undertaken conservation efforts, often in partnership with native communities, with the objectives of curtailing the degradation of natural and cultural environments, protecting ecological integrity, and safeguarding the spiritual and cultural heritage of sacred mountain sites.

Christoph Stadel, Ph.D., is emeritus professor of geography, University of Salzburg, Austria, and adjunct professor, Institute of Natural Resources, University of Manitoba, Canada. He is an expert on Andean cultural geography and human dimensions of comparative mountain geography and coauthor of the book *The Andes: A Geographical Portrait* (2015).

Notes

1. The Spanish versions of several words are provided throughout the chapter; these are words that encapsulate key concepts in Andean cultures and are used by native speakers within the region.

References

Andolina, Robert, Nina Laurie, and Sarah A. Radcliffe. 2009. *Indigenous Development in the Andes: Culture, Power, and Transnationalism.* Durham, NC: Duke University Press.

Argumedo, Alejandro and Tammy Stenner. 2008. *Association ANDES: Conserving Indigenous Biocultural Heritage in Peru.* London: International Institute for Environment and Development.

Bastien, Joseph W. 1978. "Mountain/Body Metaphor in the Andes." *Bulletin de l'Institut Français d'Etudes Andines* 7, nos. 1–2: 87–103.

———. 1985. *Mountain of the Condor: Metaphor and Ritual in an Andean Ayllu.* Long Grave, IL: Waveland Press.

Bernbaum, Edwin. 1997. *Sacred Mountains of the World.* Berkeley, Los Angeles, and London: University of California Press.

Boelens, Rutgerd and Paul H. Gelles. 2005. "Cultural Politics, Communal Resistance and Identity in Andean Irrigation Development." *Bulletin of Latin American Research* 24, no. 3: 311–327.

Cánepa, Gisela. 2008. *The Fluidity of Ethnic Identities in Peru.* Oxford: Centre for Research on Inequality, Human Security and Ethnicity.

Estermann, Josef. 1999. *Andine Philosophie. Eine interkulturelle Studie zur autochtonen andinen Weisheit.* Frankfurt: IKO-Verlag für Interkulturelle Kommunikation.

Field, Les W. 1994. "Who Are the Indians? Reconceptualizing Indigenous Identity, Resistance, and the Role of Social Science in Latin America." *Latin American Research Review* 29, no. 3: 237–248.

Gade, Daniel W. 1999. *Nature and Culture in the Andes.* Madison: University of Wisconsin Press.

García, Maria E. 2005. *Making Indigenous Citizens: Identities, Education, and Multicultural Development in Peru.* Stanford, CA: Stanford University Press.

Gelles, Paul. 2002. "Andean Culture, Indigenous Identity, and the State in Peru." In *The Politics of Ethnicity: Indigenous Peoples in Latin American States,* edited by David Maybury-Lewis, 239–266. Cambridge, MA: Harvard University Press.

Gill, Lesley. 1993. "Proper Women and City Pleasures: Gender, Class and Contested Meanings in La Paz." *American Ethnologist* 20, no. 1: 72–88.

Phélinas, Pascal. 2002. "Las Actividades Complementarias de las Explotaciones Agrícolas Peruanas." *Bulletin de l'Institut Français d'Etudes Andines* 31, no. 3: 725–750.

Radcliffe, Sarah and Nina Laurie. 2006. "Culture and Development: Taking Culture Seriously in Development for Andean Indigenous People." *Environment and Planning D: Society and Space* 24: 231–248.

Regalsky, Pablo. 1994. *La Sagesse des Andes. Une Expérience Originale dans les Communautés Andines de Bolivie.* Geneva: Fondations Simón Patiño and Pro Bolivia.
Regalsky, Pablo and Teresa Hosse. 2009. *Estrategias Campesinas Andinas de Reducción de Riesgos Climáticos.* Cochabamba: CENDA-CAFOD.
Rhoades, Robert E. 2006. *Development with Identity: Community, Culture and Sustainability in the Andes.* Wallingford, Oxfordshire, and Cambridge, MA: CABI Publishing.
Rist, Stephan. 2000. "Linking Ethics and the Market: *Campesino* Economic Strategies in the Bolivian Andes." *Mountain Research and Development* 20, no. 4: 310–315.
Salman, Ton and Annelies Zoomers, eds. 2003. *Imaging the Andes: Shifting Margins of a Marginal World.* Amsterdam: Aksant Publishers.
Sarmiento, Fausto O. 2012. *Contesting Páramo: Critical Biogeography of the Northern Andean Highlands.* Charlotte: Kona Publishers.
———. 2013a. "*Lo Andino:* Integrating Stadel's Views into the Larger Andean Identity Paradox for Sustainability." In *Forschen im Gebirge/Investigating the Mountains/Investigando las Montañas. Christoph Stadel zum 75. Geburtstag,* edited by Axel Borsdorf, 305–318. IGF-Forschungsberichte, vol. 5. Innsbruck: Verlag der Österreichischen Akademie der Wissenschaften.
———. 2013b. "Ruinas Reificadas: El Reviver Indígena, los Paisajes Culturales y la Conservación de Sitios Sagrados." *Revista Parques* 1: 1–14.
Stadel, Christoph. 1989. "The Perception of Stress by *Campesinos:* A Profile from the Ecuadorian Sierra." *Mountain Research and Development* 9, no. 1: 35–49.
———. 1991. "Environmental Stress and Sustainable Development in the Tropical Andes." *Mountain Research and Development* 11, no. 3: 213–223.
———. 2001. "'*Lo Andino:*' Andine Umwelt, Philosophie und Weisheit." *Innsbrucker Geographische Studien* 32: 143–154.
———. 2003. "Indigene Gemeinschaften im Andenraum—Tradition und Neuorientierung." *HGG-Journal (Heidelberger Geographische Gesellschaft)* 18: 75–88.
Ströbele-Gregor, Juliane. 1994. "From Indio to Mestizo … to Indio: New Indianist Movements in Bolivia." *Latin American Perspectives* 21, no. 2: 106–123.
United Nations General Assembly. 2007. *United Nations Declaration on the Rights of Indigenous Peoples: resolution / adopted by the General Assembly,* 2 October 2007, A/RES/61/295, available at: http://www.refworld.org/docid/471355a82.html [accessed 8 January 2017]
Van Wijk, Geertj. 2010. *Spirituality as Ingredient for Development of Sustainable Land Management Programs?* M.Sc. thesis. Wageningen University.
Verschuuren, Bas, Robert Wild, Jeffrey McNeely, and Gonzalo Oviedo. 2010. *Sacred Natural Sites: Conserving Nature and Culture.* London: EarthScan; Gland: IUCN.
Wilson, Fiona. 2000. "Indians and Mestizos: Identity and Urban Popular Culture in Andean Peru." *Journal of South African Studies* 26, no. 2: 239–253.
Zoomers, Annelies. 2005. "Cultura Andina, Pobreza y Bienestar: Es la Andinidad un Capital o un Obstáculo?" In *Vigencia de lo Andino en los Albores del Siglo XXI: Una Mirada desde el Perú y Bolivia,* edited by Xavier R. Lanata. Cuzco: Centro de Estudios Regionales Andinos Bartolomé de las Casas.
———. 2006. "Pro-indigenous Reforms in Bolivia: Is There an Andean Way to Escape Poverty?" *Development and Change* 37, no. 5: 1023–1046.

CHAPTER 7

National Park Service Approaches to Connecting Indigenous Cultural and Spiritual Values to Protected Places

David E. Ruppert and Charles W. Smythe

Any landscape is composed not only of what lies before our eyes but what lies within our heads.
—D.W. Meinig, "The Beholding Eye: Ten Versions of the Same Scene."

Introduction

Anthropologist Keith Basso (1996: xiii) poses a straightforward question: "What do people make of places?" This question, seemingly simple, evokes answers as varied as the people who ask it. Geographers, no newcomers to this question, have studied, classified, and mapped places for centuries to give us a visual, sometimes highly revealing, view of where we are, have been, and will be. Historians, through written documentation, describe significant human events—events that, in most all cases, draw their poignancy from the context of where they occur. Place becomes an important, often essential, element in the "story" of history. Archaeologists define and interpret places as sites and rely on analyses of unearthed artifacts to build a picture of the past life of a band, a tribe, a town, and a city—in a place.

Landscape architects, sculptors and interpreters of place, approach their work with ideas of how to shape place in ways that conform to commonly held ideas of historical integrity or visual beauty related to the design of a constructed environment—and in ways that reflect the values of the societies in which they live.[1] But it needs stating upfront that professionally trained geographers, historians, archaeologists, and architects bring their own perspective, their own sets of well-defined analytical tools to their assessment and "creation" of place. In most cases these tools are designed to impose, or overlay, a professional analytical meaning of place.

But you don't have to be a trained professional to "make places." All individuals have, and react to, a sense of place—a place where they feel at home, a place to which they attach meaning and significance from a personal, often unspoken, intuitive perspective. Place becomes both a function of space and a function of sentiment and thought where a person or a group feels comfortable or at rest or where they fit in. Distinct cultural groups, whole communities, and even nations also construct places in different ways and imbue them with shared meaning that defines common values—local or national—that are not derived from the professionals' toolkit. Understanding these cultural and ethnic vantage points, or the diversity of placemaking through the shared cultural perceptions of place found in social groups, constitutes one purpose of the National Park Service Ethnography Program in the United States.

The National Park Service

For the U.S. National Park Service (NPS), understanding the meaning of place is no small matter. National parks are established precisely because they are highly valued by a nonprofessional community; they are places that reflect shared national values, be they natural, cultural, historical, or architectural. The preservation of these places is, of course, the primary purpose of the NPS. Places are set aside and protected because they occupy not only space but also a place in the shared history or identity of the nation as whole. National parks vary in geography, natural resources, and cultural resources, and each has its own defining purpose, but together they are an amalgam of special places that resonate with, and are a reflection of, the national identity and character of the United States.

National parks are established to protect natural places, such as Yellowstone National Park with its stunning geysers and wildlife, and to preserve historic sites, such as Independence Hall as the setting for the creation of our nation. Indigenous cultural places, such as Mesa Verde in Colorado or Canyon de Chelly in Arizona, are also protected because they provide reflections of the nation's first Americans' history and culture. These places are judged to be highly significant to the nation, and the nation expends considerable resources to maintain these ancient ruins' physical structures and their historical integrity. For all these reasons, the NPS is uniquely suited to address the significance of sacred place.

The National Historic Preservation Act (1966, as Amended)

The NPS, like all federal agencies, is directed by legislation, federal regulation, and policy. Perhaps the most significant federal law to shape the agency's perspective on the importance of place is the National Historic Preservation Act (NHPA) of 1966, as amended. While some legal authorities for the protection of historic sites existed prior to this act (e.g., the Antiquities Act of 1906, 16 U.S.C. 431–433 and the Historic Sites Act of 1935, 49 Stat. 666; 16 U.S.C. 461–467), the NHPA is by far the most comprehensive historic site preservation mandate to date.

The law establishes a process to inventory and evaluate historical properties throughout the United States and assign them to a national list, the National Register of Historic Places. The very purpose of the law is to identify significant cultural places and to find management methods to protect them in the face of potential threats by federal or private action. This law was crafted to apply nationally, providing a set of broad national criteria for evaluating the historical significance of a place or site in local, state, or national historical contexts. If these criteria are met, a site can be classified as eligible for and listed on the National Register of Historic Places. Such places become subject to a certain level of consideration for preservation or protection in the face of potential negative impacts resulting from proposed federal actions.

The concept of "significance" is central to the NHPA and listing properties on the National Register of Historic Places. To qualify for the National Register, a property must meet specific criteria and embody significance as defined in the law's regulatory language. Under the NHPA, the secretary of the interior is responsible for establishing criteria for properties to be included on the National Register and criteria for national historic landmarks (NHLs). The NPS is the federal agency designated to develop and implement these criteria and maintain the National Register. The NPS has issued guidance in the form of technical bulletins for implementing these provisions of the NHPA. Whether a property is considered significant depends on its historical context.

According to NPS guidance, "The significance of a historic property can be judged and explained only when it is evaluated within its historic context" (National Park Service 2002: 7). Historical contexts are found at a variety of geographic levels or scales. The National Register recognizes local, state, and national contexts within which a property's significance is evaluated and determined. In short, "significance" and "context," while loosely defined, are at the heart of U.S. public policy addressing the preservation of historical and cultural properties. In the evaluation of significance and context, the NPS cultural resources programs contribute most to the documentation of historically and culturally important places and structures, usually (but not always) those that are located within park boundaries.

Following the passage of the NHPA and the establishment of the National Register, the significance or importance of historic sites or places was judged largely from a Euro-American perspective. Professionals wrote NHPA criteria from the standpoint of Western scholarly traditions (such as history, archaeology, and architecture). Thus it is no surprise that places put on the National Register largely reflected cultural perspectives of Western European academic disciplines or traditions. It is not hard to judge such places as Mesa Verde or Yellowstone as highly significant.

These places are recognized nationally and internationally for their importance, and such judgments of significance are rarely contested. On the other hand, smaller, less conspicuous places, such as a traditional lodge (home) of an important Indian leader, a small grove of trees, a natural spring or isolated lake, may not rise to the level of national or international significance, but they may be highly valued by a local community for religious, historical, or cultural reasons. The law allows local places to be placed on the National Register, but properties judged important or culturally significant by ethnic populations or indigenous communities were, in the past, often overlooked.

The National Park Service, Ethnography, and Sacred Geography

Because these properties were often overlooked, efforts were made by a host of cultural minority communities to press for protection of places that are considered sacred based on each group's traditional values, practices, and beliefs. In the 1970s, Native American tribes were especially active in seeking recognition and protection of important sacred and cultural places. Native American ideas of sacred places are wholly non-Western in viewpoint and more often than not combine tribal cultural history, place, religious perspective, and traditional cultural knowledge. The first serious result of these efforts was the passage of the American Indian Religious Freedom Act of 1978.

In essence, this joint resolution of Congress requires federal agencies to consult with tribal traditional leaders to determine whether federal policy or action conflicts with traditional Indian religious practices. It also requires federal agencies to provide reasonable access to public lands used for religious or ceremonial purposes. However, the real import of this resolution is the fact that it thrust non-Western cultural diversity into the arena of evaluation of site significance. Before this law, there had been a wide, unspoken gap between places considered important to Native American communities and those important to Western-trained professionals applying the standard national criteria of historical significance as outlined in NHPA regulations. By the late 1980s this gap was widely recognized and became the focus of policy, executive orders, and changes in legislation.[2]

Amendments to the NHPA of 1966 also directed attention to expanding the notion of heritage and its preservation. Amendments in 1980 directed the Department of the Interior and the American Folklife Center (a research center in the Library of Congress) to conduct a study on how to preserve intangible aspects of American cultural heritage and to encourage the continuation of diverse prehistoric, historical, ethnic, and folk cultural traditions expressed in that heritage. One of the recommendations of the ensuing report, titled *Cultural Conservation,* suggested that the NHPA could be used to more systematically address the need to protect and preserve these overlooked properties.

In response, the NPS published a set of guidelines defining a new type of significance for historical properties under the NHPA, known as "traditional cultural significance." National Register Bulletin 38 (Parker and King 1990) can be seen as a significant watershed publication, as it explains how existing National Register criteria can be applied to properties considered significant by local ethnically distinct communities (and not solely by agency professional staff). The appearance of this bulletin was critical in bringing to agencies a greater awareness of local community heritage values, something already allowed for in NHPA but often overlooked by agency academic specialists.

A historical property that has traditional cultural significance (dubbed a traditional cultural property [TCP]), "is eligible for inclusion in the National Register because of its association with the cultural practices or beliefs of a living community that are (a) rooted in that community's history, and (b) are important in maintaining the continuing cultural identity of the community" (Parker and King 1990: 1). The guidelines make the point that the existence and significance of such places may not be identified through archaeological, historical, or architectural surveys but may require in-depth interviews with persons knowledgeable about such places and other forms of research.

The bulletin gives special emphasis to Native American properties because it was found that policies and procedures of the National Register have been interpreted by federal agencies as excluding historical properties of religious significance to Native Americans. It also acknowledges that while the National Register is not the appropriate vehicle for recognizing purely intangible cultural values, the attributes that give TCPs significance are often intangible in nature.

Amendments to the law in 1992 continued to address shortcomings in the definition of "significance" for properties to be listed on the National Register. For example, the amendments formally acknowledge the need to reevaluate properties previously denied listing under earlier judgments, recognizing that the concept of significance changes over time in response to changes in current scholarship. Importantly, the new regulations accompanying these amendments (1996) give special weight to Native American perspectives on a property's cultural and religious significance—from the tribal perspective—and greatly strengthen tribal consultation provisions regarding the identification and consideration of potential adverse effects to historical properties, including sacred sites, resulting from federal actions. Since the addition of these important changes, the application of the law has recently expanded to include not only specific sites for listing but entire landscapes as well.

NPS Ethnography, Ethnic Diversity, and the Meaning of Place

In 1980, the NPS established the Ethnography Program to respond to the increased efforts of Native American communities and other ethnically distinct communities to take their unique cultural and historical perspective into consideration when land management decisions are made. Native American tribal communities have long held places to be culturally and historically significant from the perspective of their own cultural practices, values, and beliefs. Other, nonindigenous communities originating from places around the globe—including Africa, East Asia, South Asia, Central and South America, the Pacific islands, and more—also apply their own cultural values to sites and places. Like the treatment of Native American cultural perspectives, these diverse ethnic histories and viewpoints were often not considered by park managers when the NPS Ethnography Program was formed.

The program was designed to document culturally diverse views of park sites and resources and to encourage park managers to take them into consideration in their planning and management decisions. The original goal of the Ethnography Program was to conduct research with subject communities and to provide data on ethnographic resources to park managers. The NHPA is one of the principal authorities under which the NPS Ethnography Program conducts its work.

It should be noted that the NPS category of ethnographic resource goes well beyond "place." The NPS Ethnography Program views all resources as potential ethnographic resources, such as whole landscapes or viewscapes, archaeological sites, geographic features, plant communities, minerals, wildlife, and even insects. This is an important aspect of the program since its documentation work—and results—can be viewed in much the same vein as research on any natural and cultural resource. Sites and places having significance for ethnically distinct communities are identified and documented, and recommendations are made on their appropriate management within a broad resource management framework that applies to all resources for which the agency is responsible.

The difference lies in the fact that such management recommendations are related to, and informed by, the traditional cultural values, beliefs, and practices of the associated community, rather than formulated from the perspective of the professional disciplines of natural science, history, archaeology, or architecture. Simply put, the Ethnography Program focuses on the application of criteria from the local, often ethnically distinct, communities and applies local, ethnically distinct evaluations to all natural and cultural resources, including sites, places, and landscapes.

Ethnographic resources are often documented for specific and immediate management purposes. Resource management issues or conflicts in parks can arise without warning, and ethnographic studies are often designed to produce rapid results with recommendations for action to address immediate concerns. In some of these cases, site-specific information developed in collaboration with local communities provides the documentation that is needed for determinations of eligibility for listing in the National Register, which may lead to formal listing in the National Register. In either case, eligibility for or formal listing in the National Register, such sites are afforded some level of protection by triggering a formal process of project review and consideration of avoidance, minimization, or mitigation of adverse effects in the face of proposed federal actions. This process is often referred to as the "Section 106 process" after the section in the NHPA regulations that specifies the procedures to be followed.

Notes on Methods and Research Results

The NPS cultural anthropology program conducts research in-house, or it carries out research through cooperative agreements or contracts with academic institutions or private researchers. Research techniques are typical of anthropological work and include background research, ethnohistorical and archival research, participant observation, in-depth semistructured or open-ended interviews with knowledgeable individuals, and limited structured surveys. In recent years, this research work has been greatly facilitated by offices established at selected universities and staffed by NPS personnel through standing agreements with these universities. The purpose of these offices is to solicit appropriate research proposals from university faculty to conduct ethnographic research with communities having close associations with park resources and places.

The result of this work is a fairly large collection of research reports for many parks in each region. These reports provide specific park managers with contemporary resource or site use for park management planning. They also provide detailed information on cultural community interests in, and historical and cultural relationship to, sites, landscapes, and plant, animal, and mineral resources. Recent projects also include components that encourage greater accessibility to the knowledge gained from the research by parks managers, the associated communities, park visitors, and the public through public lectures and community gatherings, round table discussions, websites, exhibits, and the preparation of abridged reports presenting the research with attractive layouts and images.

These culturally important sites, especially those identified by Native American communities, can often be classified as sacred sites or sacred landscapes, and as such, both their description and location can be sensitive matters. As a consequence, agency researchers are often asked by community consultants to maintain a high level of confidentiality regarding specific aspects of the resulting reports. Maps and physical descriptions of sensitive sites are very useful in matters related to resource management; however, in some cases these are not included in reports but maintained in separate confidential files.

Case Examples

Example 1: Taos Pueblo, New Mexico

In 1989, the municipality of Taos proposed an expansion to its local airport. An Environmental Assessment by the Federal Aviation Administration (FAA) quickly followed and declared that the expansion, which included an additional runway allowing higher air traffic over the ancient pueblo, would have no environmental impact on the local community, including the pueblo structures. The pueblo immediately countered with a claim that the FAA did not adequately study the potential impacts on the pueblo structures, the lands throughout the reservation, and the traditional culture of the community. In response, the courts directed the FAA to produce a more thorough study through an environmental impact study.

Since there was some confusion regarding the significance of the tribal lands impacted, Taos Pueblo and the FAA requested assistance from the Ethnography Program in the then southwest region of the NPS. The region's ethnographer, Dr. Alexa Roberts, was joined by the region's tribal liaisons, Ed Natay and Virginia Salazar. Together they facilitated consultation with the pueblo as well as anthropological research in the mid-1990s to provide details of important cultural sites throughout reservation lands, especially those that would potentially be affected by the airport's new proposed runway and operations. The pueblo used the resulting report to negotiate with the FAA to formally increase the area of potential effect to be considered to include the entire reservation, not just the pueblo structures as originally proposed.

The study results, though shared with the NPS and the FAA, remain confidential, largely due to the sensitive nature of the cultural information involved. Prior to this study, the Southwest Region Ethnography Program also assisted the pueblo in applying to United Nations Educational, Scientific and Cultural Organization (UNESCO) to inscribe the pueblo on its World Heritage Site (WHS) list. The pueblo was listed in 1992 and remains the only WHS in the United States listed for its value to a living traditional Native American culture.

Following this study, the listing of the Pueblo on the WHS list, and the enlargement of the area under study, the focus of concern shifted from direct structural effects of vibration on the ancient buildings to the potential impact on the living traditional cultural community. The NPS and the pueblo jointly arranged for acoustical studies of aircraft overflights in those areas throughout the reservation that were identified in the ethnographic study as culturally important ceremonial sites.

Results showed that noise levels over these sites would increase due to the projected increases in air traffic and changes in aircraft approaches to the airport. However, the more important finding indicated that the frequency range of the aircraft noise directly affected the human vocal range, thus interfering with the tribal ceremonial activities that took place at these sites. All of this was new territory for the FAA, but the pueblo remained adamant that the impacts to their traditional lifeways were significant. Based on these findings, and others, the pueblo continued to press for changes in minimum overflight altitudes as well as changes to airport takeoff and landing procedures and approaches that would reduce the impact to the pueblo and to traditional cultural practices.

After twenty-two years of consultations, studies, and negotiations, all of which eventually involved the UNESCO World Heritage Committee, the FAA, the Advisory Council on Historic Preservation (ACHP), the NPS Office of International Affairs, the municipality of Taos, and a host of attorneys for all sides, the FAA and the pueblo signed a memorandum of agreement in 2012 allowing the airport expansion to move forward with innovative protection measures in place. These measures included, for example, altitude flight guidance for pilots to reduce visual and noise impacts throughout the area of potential effect.

It also established a flight-monitoring program to be managed by the pueblo with assistance from the FAA to identify private (or commercial) aircraft in violation of these guidelines or flying at very low altitudes over the pueblo. One of the original complaints the pueblo lodged was that the airport expansion would increase the number of private pilots "buzzing" the pueblo, thus causing an increase in the daily disruption of ritual life in the pueblo. What this case study highlights is the fact that through more than two decades of tribal consultation, a mutually agreeable resolution to this potential conflict was reached. It can be said with some justification that this long period of consultation, provided by the application of National Environmental Protection Act and NHPA processes, was necessary for the pueblo to clearly communicate their own views of the significance of the pueblo and pueblo lands to often skeptical federal agency officials.

More clearly stated, consultation served more as a complex cross-cultural educational process than as a communication or negotiation process. From the start, the pueblo required that it direct the research and documentation work and that it be in full control of consultation meetings reporting on the results of this work. This educational process served a vital purpose since without at least a partial understanding (by the federal agencies involved) of the shared community perspective of what makes Taos special to its people, history, culture, and contemporary tribal life, an agreeable resolution was highly unlikely.

Example 2: Cape Wind, Nantucket Sound

Nantucket Sound covers 163 square nautical miles off the southern coast of Massachusetts between Cape Cod and the islands of Martha's Vineyard and Nantucket. The Mashpee Wampanoag Tribe lives on the adjoining mainland in the town of Mashpee, and the Wampanoag of Gay Head (Acquinnah) is located on Martha's Vineyard. This case study concerns a proposal to construct 130 wind turbine generators on Horseshoe Shoal in Nantucket Sound in a 24-square-mile area in federal waters.

During consultations under Section 106 of the NHPA, the tribes expressed the view that the sound is of ongoing historical, religious, and traditional cultural significance and is thus a TCP. The Massachusetts State Historic Preservation Officer concurred with this finding and issued a formal written opinion that the sound is a TCP, providing extensive documentation based on archaeology, history, and ethnography. At this point, the Department of the Interior, Minerals Management Service (MMS), the federal agency in charge of issuing permits for the project, made a contrary determination. Because the Massachusetts State Historic Preservation Officer did not concur with the finding of the MMS, the MMS submitted a request to the Keeper of the National Register to make a determination of eligibility (DOE) for Nantucket Sound as a TCP as part of the NHPA Section 106 process for the assessment of effects of federal actions on historical properties.

Staff of the National Register conducted research and engaged in tribal consultations to compile the information needed to evaluate Nantucket Sound as a TCP. Dr. Chuck Smythe, the ethnography program manager in the northeast region, facilitated tribal consultations and contributed anthropological expertise to the evaluation process, which considered the cultural and ceremonial beliefs and practices of the two federally recognized Native American tribes. Both Wampanoag tribes identify the sound as a direct link to their ancestral origins and long-standing cultural, religious, and ceremonial practices important in maintaining the continuing cultural identity of their community.

Nantucket Sound is central to the legends that revolve around their defining cultural hero, teacher, and giant, Maushop, and his wife, Squant, whose great feats are memorialized in geographic landscape features, including the sound itself, the striking red cliffs of Gay Head, and the island of Nantucket. This narrative tradition conveys traditional Wampanoag ties to their ancestral territory, gives symbolic meaning to their relationships with the land and water, and essentially defines the sound as an integral component of their indigenous cultural landscape.

The tribes emphasize that their ancestors traversed, lived on, buried their dead on, and otherwise used the land that is now beneath the waters of Nantucket Sound in areas such as Horseshoe Shoal, which has the potential to yield significant information about the occupation of the region over the past ten thousand years. Furthermore, the sound is intrinsically important to tribal ceremonial practices that entail interactions with the sun during its emergence on the eastern horizon, as well as with certain celestial formations, which require an unobstructed view over Nantucket Sound and of the eastern view shed.

In 2010, the Keeper found that Nantucket Sound is eligible for listing in the National Register as a TCP for the reasons described above. Nantucket Sound was also determined to be a significant historical and archaeological property that has yielded and has the potential to yield additional important information about the Native American exploration and settlement of Cape Cod and the islands (Shull 2010: 2).

When the Department of the Interior MMS was unable to resolve adverse effects through consultation with the tribes, and because there were conflicting opinions issued by two Department of the Interior agencies, the office of the Secretary of the Interior took over responsibility for the process. After several government-to-government consultations with the tribes, the Secretary concluded that further consultation would not result in agreement among consulting parties on how to resolve adverse effects and terminated consultations.

At this point in the process, the ACHP submitted comments to the Secretary stating that adverse effects on historical properties (including six TCPs important to the Wampanoag tribes) cannot be avoided or mitigated and recommended that the project not be approved. The ACHP noted the presence of numerous historic properties and two national historical landmarks (including the entire island of Nantucket) within the area of potential effect, which were considered to be adversely affected by the proposal. Ultimately, the Secretary approved the project because the public benefits weighed in favor its approval.

The keeper's determination of effect significantly expanded the application of National Register criteria in comparison with earlier determinations with regard to Native American sacred places. The issue has to do with the need to identify the boundaries of historical properties for purposes of the National Register. Having clearly identifiable boundaries for historical properties has been a continuing and unresolved problem faced by Native Americans in proposing cultural and sacred sites to the National Register; this is because often an entire mountain or another large landscape feature, such as a body of water, as in this case, is considered sacred. Nantucket Sound was determined to be eligible as an integral, contributing feature of a larger, culturally significant landscape, the boundaries of which are yet to be determined.

While the keeper noted that additional documentation is necessary to define the precise boundaries of the larger landscape of which Nantucket Sound is a part, the determination answered the question for the sound itself, which was the area that prompted the request for a determination and which has a geographic boundary established by the Coast and Geodetic Survey.

This case is also groundbreaking because it relates to a large body of water. In prior cases, unofficial policy has been to discourage the consideration of natural bodies of water for eligibility determinations or nomination to the National Register (see Shull 2010: 3). Furthermore, the DOE recognized that the TCP includes both the body of water and the basin, or the floor, of the sound underneath the water, which is significant because it was a habitation area when it was possible to walk between what is now Cape Cod and the

Figure 7.1. Simulated view of proposed windmills in Nantucket Sound (photo credit: www.capewind.org. Accessed 18 July 2012)

nearby islands. This component of the property has additional significance as an archaeological property that has yielded and has the potential to yield further scientific information regarding Native American exploration and settlement of the region during historical and precontact periods.

Example 3: Medicine Wheel/Medicine Mountain National Historic Landmark, Wyoming

The Medicine Wheel NHL in Wyoming is an ancient Native American–constructed stone alignment resembling, to Western eyes, a large spoked wheel lying on its side. Located on U.S. Department of Agriculture Forest Service land near the top of Medicine Mountain in northern Wyoming at an elevation of 9,600 feet, the so-called Medicine Wheel was designated an NHL in 1970. The wheel was designated a NHL for its value as an important archaeological site; historians and archaeologists consider it an important part of a larger regional complex of stone wheels found in the upper Great Plains of the United States and Canada.

The "wheel" itself consists of a central stone cairn from which twenty-eight stone lines or "spokes" radiate to a circle of stones designating its perimeter. The structure itself has a diameter of approximately seventy-five feet, while the entire original landmark designation in 1970 encompassed 110 acres, an area that did not adequately take into consideration additional features related to the cultural significance of the site.

Native American peoples from the northern Great Plains in the United States consider the site to be a sacred place and continue to visit the site as part of periodic pilgrimages to the area. However, they have consistently claimed that the site does not stand alone; it is part of a larger complex of significant sites, all of which contribute to the cultural significance of the Medicine Wheel itself. Because of this, there have been efforts by many tribes to enlarge the limited size of the NHL to include the stone structure and most of Medicine Mountain, where it is found, thus enlarging the area to reflect more accurately the full range of features that make the Medicine Wheel significant to the tribes.

In addition, there was additional concern regarding ongoing impacts to the site. Native American people recognized early the need for additional protection of the site and formed two organizations, the Medicine Wheel Coalition and the Medicine Wheel Alliance, to work toward greater efforts to protect it. However, even with increased Forest Service efforts to protect the site, rising visitation rates by non-Indian people has, over the years, caused visible damage. Visitors sometimes tried to move rocks forming the circle, or they littered the site with unsightly debris.

Efforts by tribes, the Forest Service, and the NPS to provide greater protection of the Medicine Wheel led to two decades of research and work to justify enlarging the boundaries of the NHL. The Forest Service supported ethnographic research to gather cultural information about the wheel and its significance to Native American tribes in the region, with the purpose of gathering enough information to either support or refute tribal claims that much of the entire Medicine Mountain was a sacred place. The research findings clearly concluded that the Medicine Wheel is actually a culturally significant landscape surrounded by numerous related archaeological sites, camps, sweat lodge sites, trails, water sources, and ritual approaches that contemporary tribal peoples use to reach the Medicine Wheel for ceremonial or religious purposes. All of these places form a complex, which must be considered one integrated place. As Boggs (2003: 5/4–5/5) notes:

> In one sense the Medicine Wheel constitutes an important and particular locus of worship. In another sense, however, it also is itself integrated into the broader spatial patterns of cultural use along with other specific locales and features of Medicine Mountain. Not only is the mountain itself at the heart of traditional practice, but the very lay of the land that constitutes the Mountain becomes incorporated into and is indivisible from the form of that practice. The mountain as a geophysical feature is itself imbued with the very specific kinds of cultural meaning and cultural order that are also symbolized in the Medicine Wheel, and in the Plains Indian Sun Dance.

In sum, then, within native traditions, Medicine Mountain consists of a differentiated, culturally patterned, unified complex of natural and built features and archaeological remains and an important prehistoric trail system—all of which presently relate directly to, and can only be understood in the context of, the site's unitary, primary purpose as a major Native American spiritual and ceremonial center.

The result of extensive ethnographic study and tribal consultation was an enlargement of the NHL. This expansion encompasses much of Medicine Mountain and its associated archaeological sites, trails, and natural springs. The importance of this case example lies in the fact that, for the first time, NHL significance was determined by a set of standards (i.e., TCP criteria) designed to reflect local community values. The Medicine Wheel is an NHL and as such requires the application of criteria reflecting national significance.

When the boundaries of the Medicine Wheel were expanded to include much of Medicine Mountain as well as the wheel itself, the Forest Service was essentially applying what would normally be local TCP values. In this case, while the NPS acknowledged that the site's significance was based on local Native American cultural and historical values, it also understood that these local values were key features in a broader national context that reflect a pattern of traditional Native American religious values found throughout the continent.

The decision to expand the NHL boundaries sets a precedent because it reflects what has long been a latent understanding: that Native American sacred places are often embedded in a larger, broader context, outside of which the full significance of sites cannot be fully understood from a Native American cultural perspective. The decision to expand the NHL boundaries is tacit recognition of this fact and another step in the evolutionary changes in the application of historical and cultural criteria in the United States to better reflect indigenous religion, culture, and traditional beliefs.

These case studies provide examples of the critical use of cultural anthropology or ethnography when considering indigenous or non-Western cultural values in determinations of site significance. The NPS Ethnography Program has tried to respond to indigenous community requests that their non-Western views on place be taken into consideration. However, the program is still in its infancy, and there are many ways the professional methods and approaches of the program can benefit (and benefit from) new efforts and directions that involve programs and activities on both a national and international level.

Promising Directions—National Park Service Ethnography

Interdisciplinary Study

In the NPS, the Ethnography Program is situated in the Tribal Relations and American Cultures division within the Cultural Resources Directorate, alongside other scholarly disciplines such as history, archaeology, historical architecture, landscape architecture, and museum services. The principal activity of the Ethnography Program is to conduct research studies in applied cultural anthropology in national parks to assist in park management. The studies are designed to document different aspects of the cultural significance and traditional uses of places, landscapes, animals, plants, minerals, and other features found in national parks and park-associated communities and groups.

Studies often focus on Native Americans but also may include other ethnic and occupational groups with connections to park places and resources since the management imperatives are to (1) take into consideration the unique cultural perspectives, beliefs, and practices of these peoples in making management decisions; and (2) to educate park visitors about the cultural and historical contexts of the national park areas. The Ethnography Program also plays a leading role in educating park managers about effective consultation with Native Americans and other groups, the legal and policy frameworks for federal agency relationships with Indian tribes, the repatriation of human remains and cultural items to Indian tribes, and other issues. The Ethnography Program is the smallest of the NPS cultural resource programs and includes positions in five regional offices and about fifteen parks. Natural resource programs are situated in a parallel, but much larger, directorate, which includes biological sciences, climate change, geology, ecology, and other disciplines.

The late Canadian anthropologist Henry (Hank) Lewis once remarked that in academia the practical contributions of anthropology "rank closely with departments of religious studies and the dramatic arts, but without occupying the moral high ground of the former nor having the entertainment value of the latter" (Lewis 1992: 15). Anthropology in the NPS is similarly placed. Because the Ethnography Program often addresses natural as well as cultural resources, it finds itself wedged between these programs without having many of their respective institutional advantages, such as specific legislative authorities and mandates.

However, since ethnography addresses both natural and cultural resources, it is uniquely poised to integrate them in ways advantageous to both. Anthropology, history, and archaeology, as well as the natural resource disciplines of ecology, wildlife biology, botany, and zoology, all offer opportunities for interdisciplinary studies. Such interdisciplinary cooperation within the agency between programs has been suggested in the recent past, but it remains to be fully realized (Ruppert 2001). But interest in such work has been growing. For example, the northeast region of the NPS regularly undertakes historical research as part of its core projects to document associations with place that span hundreds of years of written history.

Starting in 2012, a Historic Resource Study will examine the history of eighteenth-century treaties with Native Americans that were signed at Fort Stanwix, New York, to be followed by an ethnographic study on their legacy from the perspectives of Native American tribes whose ancestors signed them. Also, a major interdisciplinary project is planned at Acadia National Park looking at traditional plant gathering integrated with ecological research to inventory these plant populations and identify sustainability issues.

Traditional Ecological Knowledge

Interdisciplinary work is the focus of the increasing interest in what has come to be termed traditional ecological knowledge (TEK). In many cases, indigenous peoples' use of resources over long periods of time has led to traditional techniques of resource manipulation, and a body of knowledge focusing on that use and manipulation can demonstrably foster sustainable social and ecological systems. It can also be used to shape places and landscapes in ways that address the ecological issues of conservation and sustainability (two goals often voiced by the NPS).

Early and important work in this field focused on the use of fire as a resource management technique in Australia, Canada, and California (see Lewis 1973, 1989). More recent work addresses the direct connection between the application of TEK and its environmental effects. Detailed scientific methodology has been applied to demonstrate indigenous manipulation of local plant communities to foster the sustainability of cultural and traditional uses of those resources (Blackburn and Anderson 1993; Nabhan 2000; Anderson 2006; Berkes 2008).

Recent emphasis has also focused on the use of TEK to restore ecological conditions to a previous state, one that reflects the presence of indigenous populations prior to the arrival of immigrant populations (Ruppert 2003). The NPS anthropology program could profit by expanding its interest from identifying and documenting ethnographic resources to determining how the traditional use of these resources change or sustain local ecological conditions and assisting conservation efforts in parks and other significant places (see also Painemilla et al. 2010).

New Partnerships

On a related front, partnerships between NPS resource programs and park-associated communities provide opportunities for joint research and community education. Increasingly, Native American tribal communities have increased their capacity to conduct research on their own—research that provides information on resources and traditional resource use. The NPS certainly has an opportunity to direct some of its resources to assisting tribes in this effort or to find ways to incorporate tribal findings into resource management policy and guidance. In some areas, this is referred to as "comanagement," a concept that requires greater definition and discussion. Regardless of terminology, there are increasing opportunities for future work through partnerships with constituent communities that could prove productive for all involved (Ruppert 2001; Nie 2008).

International Perspectives

Ethnography conducted in collaboration with indigenous communities offers opportunities to work more closely with international organizations and site protection efforts in other countries. In many ways, the international community has taken steps to protect traditional use areas or sacred places that move well beyond NPS efforts. Many countries are faced with the protection of sacred sites or traditional resource use while the subject communities continue to reside in them. This poses problems not often faced by the U.S. NPS, since Native American communities, with minor exceptions, no longer reside within the park boundaries. In addition, other countries are also faced with the often conflicting circumstance of balancing national and international efforts to exploit natural resources while protecting culturally significant areas that continue to be inhabited by indigenous communities. Increased communication with these governments and the NPS Ethnography Program (through the NPS Office of International Affairs) would likely result in an exchange of methods and techniques for cultural site protection that would benefit all parties.

On another front, the NPS Ethnography Program has an additional opportunity to become involved in international efforts to protect important natural and cultural places. To date, 153 nations have signed on to the UNESCO World Heritage Convention, first proposed by the United States in 1972. This program inscribes significant sites and places on the World Heritage List and, once listed, requires member states to report periodically on continued condition or integrity. Since President Nixon signed the convention in 1972, the United States has successfully nominated and placed twenty-two sites on this international list of protected places, many of them national parks.

Since 725 of the 936 sites listed around the world are cultural sites and involve indigenous populations, the Ethnography Program may have much to learn from other countries' methods and techniques of working with these communities to foster conservation and sustainability as they relate to the identification and protection of WHSs. On the other hand, given the long history of the NPS, the Ethnography Program could provide useful insights for other member states. In either case, increased involvement in the WHS program would benefit all parties. The NPS's Office of International Affairs serves as the nation's steward of this UNESCO program.[3]

International Union for Conservation of Nature Natural Sacred Sites Guidelines

The International Union for Conservation of Nature (IUCN), in collaboration with UNESCO's Man and the Biosphere Program, has taken a lead in the protection of sacred natural places. In 2008, an IUCN taskforce published *Guidelines for Protected Area Managers* (Wild and McLeod 2008), which emphasizes the management and protection of areas sacred to human communities. The publication provides sacred site principles and goals designed to be generally applicable, with a focus on working with indigenous peoples, traditional cultural knowledge, and the integration of natural and cultural research. The guidelines are comprehensive and address in detail principles related to working with indigenous communities and documenting cultural information. Since these principles and guidelines address natural areas considered sacred by constituent communities, they may provide useful elements to integrate into the NPS's Ethnography Program.

Intangible Cultural Resources

Another UNESCO convention, "The Convention on the Safeguarding of the Intangible Cultural Heritage," was adopted by many member states in 2003. Since its adoption, 127 member states have accepted and ratified the convention. Basically, the convention calls for the preservation and protection of what it terms "intangible" cultural resources and considerably expands the notion of what cultural traditions need protecting. As the UNESCO website states:

> Cultural heritage does not end at monuments and collections of objects. It also includes traditions or living expressions inherited from our ancestors and passed on to our descendants, such as oral traditions, performing arts, social practices, rituals, festive events, knowledge and practices concerning

> nature and the universe or the knowledge and skills to produce traditional crafts. (UNESCO 2016)

The United States has not ratified this convention (and is not likely to do so in the near future). However, discussions at the international level continue, and the subject matter seems particularly consistent with ethnographic resource goals.

It would be appropriate to revisit the issues of intangible cultural resources that were discussed in the congressionally mandated study, titled *Cultural Conservation* (Loomis 1983), since the only concrete result of that study seems to be the concept of traditional cultural significance attached to material properties in the National Register (described above). This review must proceed from the expansive vantage point from which it was commissioned, which referred to intangible cultural heritage such as arts, skills, folklife, and folkways and the need to preserve, conserve, and encourage the continuation of the diverse traditional, prehistoric, ethnic, and folk cultural traditions that underlie and are a living expression of our American heritage (NHPA 1980; Loomis 1983). Increased interagency cooperation would likely result from an expanded perspective since a number of federal agencies focus on the preservation of these traditional cultural resources.

Distribution of Research Results

Completed ethnographic reports often end up as "gray" literature that has limited distribution and collects dust on NPS office shelves. Or the researchers use the results to produce their own published works with an emphasis on their specific academic interests. While the cost of traditional publishing limits the distribution and subsequent use of these ethnographic reports, electronic distribution of reports is becoming more common, and efforts to distribute studies in this manner would prove beneficial. Increasingly, publishing involves more than print and paper. Electronic versions of books and reports need not simply replicate older forms of publication.

Modern publishing techniques allow the downloading of entire reports, or entire volumes, with interactive content, for example, links to related websites or embedded video content. Redirecting research products to take advantage of this technology would not only help reach a wider readership (including access by communities who are the subject of study) but would also provide highly useful products for interpretation in parks. To respond to the need to make its reports widely available to parks, partners, and the public, the NPS has implemented a web-based solution named Integrated Resource Management Applications, or IRMA, which is the beginning of a one-stop source for resource-related data and information, including reports and other documents, data sets, maps, images, links, and more (https://irma.nps.gov).

Of course, video is already used in ethnographic interviewing. In limited cases, ethnographic research in parks has resulted in video recordings of interviews with park-associated community members. While these recordings are often an artifact of the research process, they can also become a product apart from the more lengthy report—and, again, useful for interpretation and/or education programs. For example, high-quality video recordings of Apache elders at the Chiricahua National Monument were edited and subsequently used in the visitor center to illustrate the importance of the monument to Native peoples.

Conclusion

The United States can rightfully take credit for the modern idea of a National Park system, a system born in the middle of the nineteenth century and since exported, in one form or another, to many other parts of the world. The earliest parks were essentially places set aside to avoid the onslaught of westward expansion and the results of ecologically damaging resource exploitation. The early promoters of parks were visionaries who witnessed the ongoing destruction of majestic mountain vistas and the decimation of vast herds of wild game. The need for some measure of protection was evident to most who witnessed the rush to settle the western half of the continent. Simply put, but often politically difficult to achieve, parks were to be places wherein the wonders of the natural world would be protected for their own sake from ongoing development, as well as for the enjoyment of future generations.

These early environmentalists were inspired by an honest and ardent desire to preserve the natural landscapes of the west, but they were also blind to the perspectives of cultural communities far removed from their own. In short, this early push to protect the environment left little room for the indigenous populations who considered these protected places home—places to live, to raise children, to worship, and to provide their for subsistence needs. Overall, human occupation and consumptive resource use were viewed as incompatible with the preservation ethic; humans were viewed as a destructive force in protected areas, and many who lived within their boundaries were removed. Indigenous communities were actively discouraged from living in or using these places, or they were simply removed and settled on the reservation system. Removal was especially difficult for Native American populations since the newly established off-limits parks and preserves not only were homelands but were places where they hunted, gathered, buried their ancestors, and worshiped at sacred sites.

To make matters worse, this general attitude and the resulting policies of removal and forced enculturation continued into the twentieth century and were coupled with a general discrimination against Native American people. Federal and state governments relentlessly promoted efforts to assimilate indigenous communities into the general population. An important element of these efforts was to openly discourage or forbid Native Americans from engaging in their traditional religious practices, which were irrevocably linked to their homeland, natural resources, entire landscapes, and sacred sites. It should be no surprise to anyone that Native American people speak reverently of places, mountains, rivers, springs, or other natural features. These places are linked to Native American people through eons of personal and community association, through an intimate knowledge of place and the role these places play in their collective cultural and spiritual lives and traditions.

On a more positive note, the past three decades have witnessed a shift in attitude and policy regarding Native American perspectives on the importance of place. Federal agencies have reflected this change through changes in regulation, policy, and general guidance and by directing personnel to consult with Native American tribes on a regular basis. The Ethnography Program within the NPS is one important example of this shift in attitude and policy, which is further reflected in a recent reorganization of these research and consultative functions within the NPS under the Tribal Relations and American Cultures division.

However, agencies need not take too much credit. The changes in policies and programs have been largely, if not wholly, due to the efforts of Native American tribes themselves. Tirelessly campaigning for and supporting legislation such as the American Indian Religious Freedom Act of 1978, the Native American Graves Protection and Repatriation Act of 1990, and the 1992 changes to the NHPA of 1966, tribes have sought to institutionalize processes that force agencies to attend to the unique views held by tribes and to put in place a measure of protection for their ancestors' remains, their traditional homelands, and places and resources they consider to have cultural and religious importance.

So, "What do people make of places?" Well, the simple answer is "a lot." And, given the nation's vast cultural diversity, the range of answers can seem daunting. But the question, while complex, remains important and relevant. As members of culturally distinct communities continue to insist that their perspectives matter and should be heard, the need to study and document these perspective only increases—for historical, legal, social, political, cultural, ecological, and moral reasons. Asking this, in essence anthropological, question risks responses that range from the local to the global, and answers reveal nothing less than the holistic nature of culturally unique worlds. Again, returning to Basso, when asking this question of any community, he suggests we need to be ready for the answers they supply. Their answers, he writes:

> should not be taken lightly, for what people make of their places is closely connected to what they make of themselves as members of society and inhabitants of the earth, and while the two activities are separable in principle, they are deeply joined in practice. If place-making is a way of constructing the past, a venerable means of doing human history, it is also a way of constructing social traditions and, in the process, personal and social identities. We are, in a sense, the place-worlds we imagine. (Basso 1996: 7)

The NPS Ethnography Program continues to ask this question and continually strives to make the answers it receives relevant to land management agencies, to specific site managers, to the constituent peoples involved, as well as to the general scientific community and the public at large.

David E. Ruppert, Ph.D., is a research professor in the Department of Anthropology, University of Colorado Denver, Colorado, and an expert in ethnography, community development and protected area management, and the return of Indian human remains to tribes of origin.

Charles W. Smythe, Ph.D., is director of the Department of Culture and History at Sealaska Heritage Institute, Juneau, Alaska, an Alaska Native nonprofit cultural and educational organization whose mission is to perpetuate and enhance Tlingit, Haida, and Tsimshian cultures. He is an expert in the ethnography of protected areas.

Notes

1. Interest in geography and culturally significant places is certainly not new. What is new is the interest in the significance of places from the perspective of individuals and social groups that define such places and express their relationship to them, including non-Western cultural communities (see Thornton 2008: 4–7). Anthropology has always sought to provide geographical and/or ecological context for the cultural groups they study, but this context was one normally provided by the anthropologists, not the community members under study. Place was often seen as unproblematic, a setting for the group's social structure or economy. This has, of course, changed. Meinig's (1976) work provided early insights into the importance of place from a cultural per-

spective. Deward Walker's (1991) work on sacred geography in the 1990s foreshadowed an increased interest in the intersection of religion, geography, and culture. Henry Lewis's (1989) earlier work in Australia and Canada provides a similar early focus and added the significance of ecological consequences of indigenous use of places and resources. Fred Myers' work (1986) on concepts of self and place in Aboriginal Australia and his other papers on this topic provide a detailed anthropological investigation of these issues. In the United States, much of this evolving interest is linked to the increased political awareness of Native Americans—who have for the past few decades demanded that attention be given to their religious freedoms—and how the exercise of these freedoms are intimately connected to sites, places, and landscapes (e.g., the passage of the American Indian Religious Freedom Act of 1978).

2. Executive orders in the 1990s directed agencies to consult with Native American tribes on proposed changes in policy and agency-generated regulations that affected tribal issues. Orders were also issued to identify and protect Native American sacred sites in a manner that does not conflict with other federal law (see Executive Order 13175; Executive Order 13007). Amendments to the NHPA in 1992 provided specific legislative language directing agencies to consult with tribes on potential impacts to places that are eligible for listing in the National Register of Historic Places because of their cultural and religious significance to Native American tribes (NHPA, Sections 110 and 106).
3. The reader is referred to the following website for further information: https://www.nps.gov/subjects/internationalcooperation/worldheritage.htm.

References

Anderson, M. Kat. 2006. *Tending the Wild: Native American Knowledge and Management of California National Resources*. Berkeley and Los Angeles: University of California Press. American Indian Religious Freedom Act of 1978. P.L. 95-341, 92 Stat. 469. 3 April 1978.

Basso, Keith H. 1996. *Wisdom Sits in Places: Landscape and Language among the Western Apache*. Albuquerque: University of New Mexico Press.

Berkes, Fikret. 2008. *Sacred Ecology*, 2nd ed. New York: Routledge.

Blackburn, Thomas C. and Kat Anderson. 1993. *Before the Wilderness: Environmental Management by Native Californians*. Menlo Park, CA: Ballena Press.

Boggs, James. 2003. *Medicine Mountain Ethnographic Report*. Denver: U.S.D.A. Forest Service.

Chapin, Mac, Zachary Lamb, and Bill Threlkeld. 2005. "Mapping Indigenous Lands." *Annual Review of Anthropology* 34: 619–638.

Executive Order No. 13007: Indian Sacred Sites. 61 FR 26771, 96-13597. 5 May 1996.

Executive Order No. 13175: Consultation and Coordination with Indian Tribes. FR 65218, 00–29003. 6 November 2000.

Lewis, Henry T. 1973. *Patterns of Indian Burning in California*. Ramona, CA: Ballena Press.

———. 1989. "Ecological and Technological Knowledge of Fire: Aborigines versus Park Managers in Northern Australia." *American Anthropologist* 91: 940–961.

———. 1992. "The Technology and Ecology of Nature's Custodians: Anthropological Perspectives on Aborigines and National Parks." In *Aboriginal Involvement in Parks and Protected Areas,* edited by Jim Birchhead, Terry DeLacy, and Laurajane Smith. Australian Institute of Aboriginal and Torres Strait Islander Studies. Canberra: Aboriginal Studies Press.

Loomis, Osmond H., coordinator. 1983. *Cultural Conservation: The Protection of Cultural Heritage in the United States: A Study of the American Folklife Center.* Report number 10. Washington, D.C.: Library of Congress in cooperation with the National Park Service.

Meinig, D.W. 1976. "The Beholding Eye: Ten Versions of the Same Scene." *Landscape Architecture* 66: 47–54.

Myers, Fred R. 1986. *Pintupi Country, Pintupi Self: Sentiment, Place, and Politics among Western Desert Aborigines.* Washington, D.C.: Smithsonian Institution Press.

Nabhan, Gary P. 2000. "Native American Management and Conservation of Biodiversity in the Sonoran Desert Bioregion." In *Biodiversity and Native America,* edited by Paul E. Minnis and Wayne J. Elisens, 29–43. Norman: University of Oklahoma Press.

Nabokov, Peter. 2006. *Where the Lightning Strikes: The Lives of American Indian Sacred Places.* New York: Viking Press.

National Historic Preservation Act of 1966 (NHPA), as amended. 1980. Public Law 89–665; 54 U.S.C. 300101 et seq. Title III, Section 502.

National Park Service. 2002. *How to Apply the National Register Criteria for Evaluation.* National Register Bulletin 15. Washington, D.C.: National Park Service. Available at: http://www.nps.gov/nr/publications/bulletins/nrb15 (accessed 10 March 2012).

Nie, Martin. 2008. "The Use of Co-Management and Protected Land-Use Designations to Protect Tribal Cultural Resources and Reserved Treaty Rights on Federal Lands." *Natural Resources Journal* 28: 585–647.

Painemilla, Kristin Walter, Anthony B. Rylands, Alisa Woofter, and Cassie Hughes. 2010. *Indigenous Peoples and Conservation: From Rights to Resource Management.* Arlington, VA: Conservation International.

Parker, Patricia and Tom King. 1990. *Guidelines for Evaluating and Documenting Traditional Cultural Properties.* National Register Bulletin 38. Washington, D.C.: National Park Service.

Ruppert, David. 2001. "New Tribe/Park Partnerships: The Use of Ethnobotany in the Restoration of Indigenous Landscapes." *Cultural Resource Management* 24, no. 5: 45–49.

———. 2003. "Building Partnerships between American Indian Tribes and the National Park Service." *Ecological Restoration* 21, no. 4: 261–263.

Shull, Carol D. 2010. *Determination of Eligibility, Nantucket Sound.* Washington, D.C.: National Park Service.

Smythe, Charles W. 2009. "The National Register Framework for Protecting Cultural Heritage Places." *George Wright Forum* 26, no. 1: 14–25.

Thornton, Thomas F. 2008. *Being and Place among the Tlingit*. Seattle: University of Washington Press in association with Sealaska Heritage Institute, Juneau.
United Nations Educational, Scientific and Cultural Organization [UNESCO]. 2016. "What Is Intangible Cultural Heritage?" Available at: http://www.unesco.org/culture/ich/index.php?pg=00002 (accessed 7 December 2016).
Walker, Deward E., Jr. 1991. "Protection of American Indian Sacred Geography." In *Handbook of American Indian Religious Freedom*. Edited by Christopher Vecsey, 100-115. New York: The Crossroads Publishing Co.
Wild, Robert and Christopher McLeod, eds. 2008. *Sacred Natural Sites: Guidelines for Protected Area Managers*. Best Practice Protected Area Guidelines Series No. 16. Gland, Switzerland: International Union for Conservation of Nature.

Part 3

Case Studies

Introduction to Part 3

Case Studies

Fausto Sarmiento and Sarah Hitchner

More actively than any other hybrid disciplinary trend, the narrative produced by the joining of sacred natural sites and indigenous revival provides abundant opportunities for scholars, social and environmental activists, and others to more fully explore the "essence of place" as a way to fuse science and society (Odum and Sarmiento 1998), intellect and spirit (Kumar 2005), and native and nonnative ideologies (Nayar 2002) about nature and culture as a gradient of interactions instead of a dialectic binary. We believe that this is an important about-face required to understand the transitions from the old story of arborescent hegemony regarding resource exploitation to a new story of symbiotic awakening, rhyzomic interactions, and resource conservation.

An important element in our decision to collate the following chapters is that they exemplify the paradigm of "biocultural landscape" in which ongoing interactions between humans and their environments have determined or "manufactured" the conditions of the present landscape (Scheiber and Zedeño 2015). Of necessity, then, the anthropological lens of the "situated knowledge" complements the geographical imperative of "historicity," and these disciplinary perspectives are woven together in the following case studies, which present realities of unique phenomena that are uniquely situated in a particular locale (see Nazarea 1999). At the same time, these chapters offer readers the opportunity to learn from experience and see the similarities and differences among the various contexts throughout the region of the Americas. Understanding these diverse contexts also helps to generalize the application of theoretical frameworks or management experiences regarding intangible heritage and indigenous revival.

We provide case examples, the study of which could be applied in different geographies, or even in different sociologies. We selected exemplary cases in order to emphasize the importance of indigenous revival within the Americas. The Cherokee, for instance, show respect for their ancestral roots and beliefs while flourishing amid development pressures in the Appalachians (Chapter 8 by Steere). Chapter 9 (Bastida Muñoz and Encina) explores the applicability of the biocultural landscape paradigm and considers the multiplicity of ethnic and indigenous groups in Mexico with various, and sometimes conflicting, views and claims on landscapes.

In Chapter 10, Viteri O. presents a case in which the political sphere influences conservation management and sacred sites in the tropical Andes of Ecuador, noting the governmentality associated with both the sacred and the indigenous. Finally, Roca Alcazar (Chapter 11) describes an example from the Amazon region that links the ethnobotanical knowledge of the Peruvian traditional selva inhabitants with development pressure and sustainability concerns; he questions whether the sacred sites of the headwaters of the Amazon can be saved from the encroachment of mining and timber concessions and whether indigenous cultures can be custodians of the incredible biodiversity of the Andean piedmont.

References

Kumar, Satish. 2005. "The Spiritual Imperative." *Resurgence and Ecologist* 229. Available at: http://www.resurgence.org/magazine/article653-SPIRITUAL-IMPERATIVE.html (accessed 13 September 2016).

Nayar, Bhaskaran. 2002. "Ideological Binarism in the Identities of Native and Non-native English Speakers." In *Us and Others: Social Identities across Languages, Discourses and Cultures,* edited by Anna Duszak, 463–479. Pragmatics and Beyond New Series. Amsterdam: John Benjamins.

Nazarea, Virginia D., ed. 1999. *Ethnoecology: Situated Knowledge/Located Lives.* Tucson: University of Arizona Press.

Odum, Eugene and Fausto O. Sarmiento. 1998. *La Ecología: El Puente entre Ciencia y Sociedad.* Mexico City: Editorial McGraw-Hill Interamericana de México.

Scheiber, Laura L. and Maria Nieves Zedeño, eds. 2015. *Engineering Mountain Landscapes: An Anthropology of Social Investment.* Salt Lake City: University of Utah Press.

Chapter 8

Collaborative Archaeology as a Tool for Preserving Sacred Sites in the Cherokee Heartland

Benjamin A. Steere

Introduction

Archaeology has the potential to play an important role in the preservation of sacred sites in North America. In certain cases, locations that are thought to be sacred by Native American communities can be identified using archaeological methods. This is true for many sites considered sacred by the Eastern Band of Cherokee Indians in western North Carolina. The sacred Cherokee landscape is extremely complex; it includes both natural features (e.g., certain waterfalls and mountains) and cultural features created by the Cherokee and their ancestors (Mooney 1900; Perdue 1998; Duncan and Riggs 2003). Some of the sacred sites that can be identified archaeologically include villages that contain ancestral Cherokee graves, the remains of historic period Cherokee townhouses, prehistoric mounds, and sites associated with petroglyphs and other rock art (Riggs and Shumate 2003; Diaz-Granados 2004; Rodning 2010).

In recent years, the Eastern Band of Cherokee Indians has collaborated with archaeologists to develop projects aimed at understanding and preserving sacred Cherokee sites and enhancing Cherokee cultural identity (Cooper 2009). This collaboration is representative of a broader movement referred to as indigenous archaeology, which is most concisely defined as archaeology that is by, for, and about indigenous communities (Colwell-Chanthaphonh et al. 2010). Working in cooperation with professional and academic archaeologists, the Eastern Band of Cherokee has taken an active role in developing the research design of archaeological studies for cultural resource management projects and academic endeavors. Such projects make broad contributions to archaeological knowledge while also respecting traditional Cherokee beliefs about the treatment of sacred places, graves, and ceremonial objects.

The Western North Carolina Mounds and Towns Project, a collaborative archaeological research project initiated by the Tribal Historic Preservation Office of the Eastern Band of Cherokee Indians and the Coweeta Long Term Ecological Research Program at the University of Georgia, is one such effort. Western North Carolina once contained many mounds, monumental earthen structures built by Native Americans from approximately AD 200 until the historic period (Dickens 1976; Keel 1976; Ward and Davis 1999; Rodning 2009; Kimball, Whyte, and Crites 2010, 2013). These mounds are sacred places on the Cherokee cultural landscape, but many of these sites have been damaged by looting, development, and modern agriculture, and in some cases their locations have been forgotten.

The primary goal of the Western North Carolina Mounds and Towns Project is to create a map and database documenting all the prehistoric and historic period mound sites in the eleven westernmost counties of North Carolina. This project is ongoing, but it has already produced important new information for preserving sacred Cherokee sites and revitalizing Cherokee culture, new data for generating a broader understanding of Cherokee historical geography, and new opportunities for collaborative archaeological research (see Steere 2015).

This chapter describes the development and initial results of the Western North Carolina Mounds and Towns Project and discusses its contributions to the broader fields of sacred sites preservation, indigenous revival, and indigenous archaeology. Following a brief cultural and historical introduction to the background of western North Carolina, I present the initial results of the project in terms of their potential for protecting sacred sites and contributing to cultural revitalization efforts for the Eastern Band of Cherokee Indians. In closing I suggest that this project can serve as a model for archaeological research and preservation efforts that are by, for, and about indigenous communities.

Cultural and Historical Background

Western North Carolina is the ancestral homeland of the Cherokee people. Today, about 60 percent of the thirteen thousand enrolled members of the Eastern Band of Cherokee Indians live on the Qualla Boundary, an approximately 57,000-acre reservation adjacent to the Great Smoky Mountains National Park, which includes the town of Cherokee, North Carolina. This roughly 100-square-mile area represents a small fraction of the approximately 125,000-square-mile territory the Cherokees may have controlled in the early eighteenth century, based on archaeological evidence, early written accounts, and Cherokee oral history (Finger 1984; Duncan and Riggs 2003; Gragson and Bolstad 2007). Population estimates for the size of the Cherokee nation in the mid to late eighteenth century fall to around thirty-six thousand people living in approximately sixty towns in South Carolina, Georgia, North Carolina, and Tennessee (Smith 1979; Duncan and Riggs 2003; Gragson and Bolstad 2007).

The members of today's Eastern Band are descendants of a group of approximately one thousand Cherokees who survived late eighteenth-century wars with European and American forces and multiple smallpox epidemics and then resisted removal in 1838. By the early twentieth century, these survivors had established the Eastern Band of Cherokee Indians as a federally recognized tribe and sovereign nation, with their lands held in trust by the federal government (Duncan and Riggs 2003).

The study area for this project includes the eleven westernmost counties of North Carolina, which were home to the Valley, Middle, and Out Towns of the Cherokee in the eighteenth century (Smith 1979; Boulware 2011). The eleven counties fall within the Southern Blue Ridge Province of the Appalachian Mountains (Fenneman 1938), and the terrain is dominated by steep mountains, sharp ridge tops, and narrow valleys. The major river drainages in the study area, from east to west, are the French

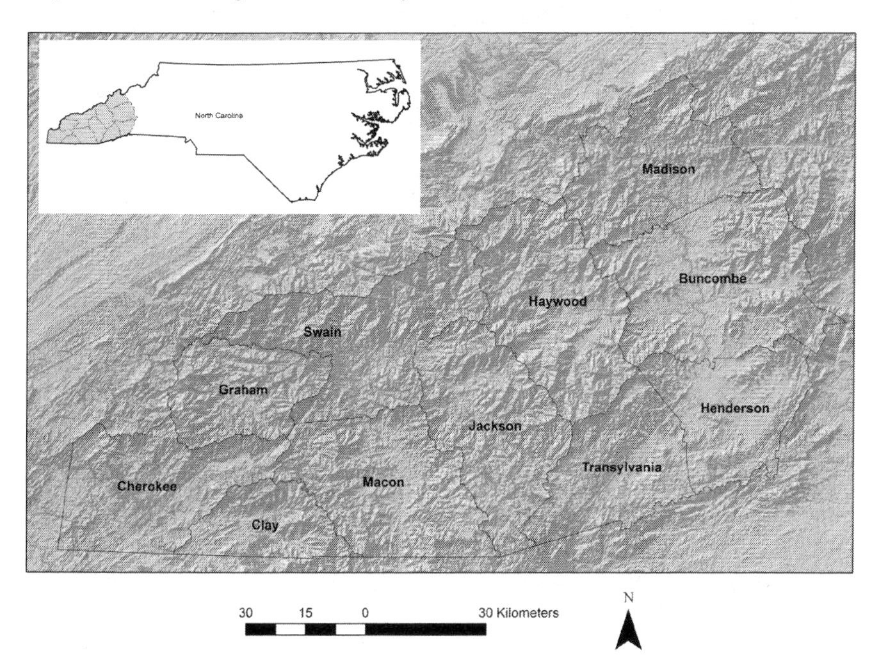

Figure 8.1. Map of the Study Area in Western North Carolina

Broad, Pigeon, Tuckasegee, Little Tennessee, and Hiwassee. This area is generally considered to be the Cherokee "heartland" (see Gragson and Bolstad 2007), and it includes the mother town of Kituwah, which, according to oral tradition, is the Cherokee place of origin (Mooney 1900).

In western North Carolina and the surrounding Southern Appalachian region, Native American communities began building mounds during the Middle Woodland period, around AD 200 (Keel 1976; Kimball, Whyte, and Crites 2010, 2013). The best documented Woodland period mound sites in the Cherokee heartland include the Connestee phase Mound No. 2 at the Garden Creek site (Keel 1976; Wright 2013, 2014) and the Biltmore Mound, located on the grounds of the Biltmore estate (Kimball and Shumate 2003; Kimball, Whyte, and Crites 2010, 2013). Both of these mounds apparently served as low platforms for ceremonial activities and contain artifacts typically associated with Middle Woodland period ceremonial and exchange systems (Keel 1976; Kimball, Whyte, and Crites 2010, 2013; Wright 2013).

During the Mississippian period (AD 1000–1500), indigenous people in western North Carolina, following broader cultural and demographic trends in the Southeast, began practicing intensive maize agriculture and living year-round in permanent, nucleated villages (Dickens 1976; Muller 1997; Smith 1992). As seen in adjacent regions in the Southeast, such as northern Georgia and eastern Tennessee, the transition from the Woodland to the Mississippian period in western North Carolina is marked by a change in the style and function of mounds (Hally and Mainfort 2004).

Across the Southeast, mounds constructed during the Mississippian period served as platforms for elite residences and temple buildings containing sacred objects (Lindauer and Blitz 1997; Milner 2004). Mississippian period platform mounds range in size from modest structures less than one meter in height to the remarkable Monk's Mound at Cahokia, which stands thirty meters tall (Anderson 1994; Lindauer and Blitz 1997). Villages with mounds were the social and political centers of native polities, and they would have been the locus for important political and ceremonial activities (Beck 2003; Hally 2006).

In western North Carolina, Mississippian period platform mounds appear to have served similar functions. However, it is important to note that mound sites in the Cherokee heartland were generally not as large or elaborate as mound sites in neighboring regions (Hudson 1997). Large Mississippian period communities like Etowah, Moundville, and Cahokia contained multiple platform mounds and appear to have been the administrative centers of settlement systems with two or more hierarchical levels of political organization (Beck 2003; Hally 2006; King 2003; Knight 2010).

In contrast, Mississippian period central places in western North Carolina contained single platform mounds. Based on the available archaeological data, there is no clear evidence for multiple levels of political organization within settlement systems in the region. Instead, western North Carolina may have been marked by settlement systems in which a single mound site served as the political center for several surrounding communities. Similar Mississippian polities are well documented in nearby northern Georgia (Anderson 1994; Hally 1996, 2006).

After about AD 1600 and into the late eighteenth century, townhouses, large public structures measuring roughly ten to twenty meters in diameter, replaced mounds as the central public architecture of native villages (Rodning 2009, 2010). In some cases, Cherokee communities constructed townhouses on top of existing platform mounds built centuries earlier. Townhouses were rebuilt in place over time, and gradually a low mound would be formed, creating an elevated base for new townhouses. The Kituwah Mound, shown here in a photograph from 1937, is an example of such a mound (Figure 8.2). A noninvasive geophysical survey conducted in 2001 revealed that the mound at Kituwah contains multiple stages of townhouse construction (Riggs and Shumate 2003). These superimposed stages now appear as a low mound, flattened and spread out by decades of plowing by local landowners.

During the midsixteenth century, Europeans began to explore the edges of the Cherokee world. Juan Pardo and a group of Spanish soldiers established the short-lived Fort San Juan at the ancestral Catawba town of Jo-

Figure 8.2. The Kituwah Mound in 1937 (photo credit: National Parks Service)

ara, east of the Cherokee towns, and accounts by Pardo's soldiers indicate that Cherokee from Kituwah and Nikwasi visited the community (Booker, Hudson, and Rankin 1992). Cherokee had sustained contact with Europeans by the late 1600s, and townhouses served as important meeting places with traders, soldiers, and other delegates from Europe and the newly formed American colonies. The eighteenth-century accounts of William Bartram (1928 [1775]) and Timberlake (King 2007) are particularly descriptive and, when combined with insights from Cherokee oral history, have helped archaeologists interpret the archaeological remains of townhouses (Rodning 2009: 631–634).

The Cherokee townhouse at the Coweeta Creek site is one of the best preserved and archaeologically understood examples of these structures (Rodning 2002, 2010, 2015). This large public building had at least six successive stages and was used from the 1600s to the late 1700s (Rodning 2010). In contrast to platform mounds, which physically and symbolically elevated the chief above other community members, townhouses were public structures that likely functioned as an architectural symbol of the Cherokee town, emphasizing the importance of community identity over individual leadership (Rodning 2010).

During the historic period, a sacred fire was kept burning in Cherokee townhouses, and once a year, all the hearths in the village were extinguished and then ceremonially rekindled from this sacred fire. This practice may date to the Mississippian period. Based on traditional Cherokee beliefs, sacred fires continue to burn at places like Kituwah (Mooney 1900; Duncan and Riggs 2003). Cherokee myths also suggest that mounds were the home of the *Nunnehi,* immortal spirit people, and that mounds and townhouses are symbolically associated with mountains (Mooney 1900; Rodning 2009, 2010). Thus, in addition to serving as hubs for social and political activities, townhouses created a link between the built environment and sacred aspects of the natural landscape.

For nearly two thousand years, mounds have been important places on the physical and cultural landscape of western North Carolina (Mooney 1900; Duncan and Riggs 2003; Rodning 2009, 2010). Based on traditional Cherokee beliefs, sacred fires still burn today at places like the Kituwah Mound (Mooney 1900; Duncan and Riggs 2003; Riggs and Shumate 2003). Today, the Kituwah Mound is still used by Cherokee as a meeting place for certain cultural events, and the site has been carefully protected from encroaching development (Duncan and Riggs 2003). The nature and function of mound building changed significantly from the Woodland to the historic period, but the long tradition of building mounds at important central places in western North Carolina speaks to the lasting cultural importance of this architectural practice.

Previous Archaeological Research in Western North Carolina

Despite the obvious importance of these sites, and despite a rich history of archaeological research in the area, we still know relatively little about many of the mounds in the region. As in other parts of the Southeast, this is primarily the result of antiquarian excavations and other processes of site destruction. Additionally, some traditional Cherokee knowledge about mound locations has been lost, a result of the forced removal westward on the Trail of Tears in 1838 and the forced acculturation of children in American boarding schools in the early twentieth century (Perdue and Green 2001).

As in much of the eastern United States, the earliest archaeological studies in western North Carolina were sponsored by museums. From the 1870s through the early 1930s, archaeological fieldwork was carried out primarily by museum personnel and local hired laborers, with the goal of obtaining artifacts for display (Ward and Davis 1999). The first early excavations in western North Carolina were sponsored by the Valentine Museum of Richmond, Virginia. In the late 1870s and early 1880s, Mann S. Valentine and his sons, E.E. and B.B. Valentine, directed expeditions in Haywood, Jackson, Cherokee, and Swain counties, sometimes with the help of local residents, including A.J. Osborne of Haywood County and R.D. McCombs of Cherokee County (Valentine, Valentine, and Valentine 1889; Ward and Davis 1999).

The Valentines and their associates "opened" the Peachtree Mound, the Garden Creek Mound No. 2, the Wells Mound (one of a group of mounds on the West Fork Pigeon River, west of Waynesville), the Jasper Allen Mound (located on Scotts Creek, east of Sylva), the Kituwah Mound, the Nununyi Mound, the Birdtown Mound, and the Cullowhee Mound (Valentine, Valentine, and Valentine 1889; Ward and Davis 1999). These investigations were not carried out to modern standards and were highly destructive.

In the 1880s, the Smithsonian Institution carried out a large-scale survey of mounds in eastern North America and identified approximately forty mounds in western North Carolina (Thomas 1891, 1894). The results of this work were published in the annual reports of the Bureau of American Ethnology (Thomas 1891, 1894) and are also mentioned in at least one Peabody Museum report (Putnam 1884). These reports were adequate for their time but generally provide little more than an approximate location for each recorded mound and a brief description of the stratigraphy and contents of excavated mounds.

In May of 1913, Robert Dewar Wainwright, a retired captain of the U.S. Marine Corps and an amateur archaeologist, carried out surface collections and excavations at several mound and village sites in western North Carolina, including the Donnaha site, the Cullowhee Mound, the Andrews Mound, and the Kituwah Mound (Wainwright 1913, 1914a, 1914b). Wainwright spent his summers "hunting for camp sites, exploring mounds and looking for specimens of stone art" (Wainwright 1913: 111). Wainwright published a written account of his travels, "A Summer's Archaeological Research," in an obscure journal, *The Archaeological Bulletin*, which has only recently received scholarly attention (see Steere, Webb, and Idol 2012).

The next excavations in western North Carolina were carried out in Haywood County by the Museum of the American Indian/Heye Foundation (Heye 1919). In 1915, George Heye directed excavations at the Garden Creek sites near Canton, North Carolina, and excavated a mound on the Singleton property near Bethel, North Carolina (Heye 1919). Heye's 1919 report of his work in Haywood County contains more detail than most early reports but still falls short of the standards for archaeological reporting established during the 1930s.

In 1926, Charles O. Turbyfill, a Waynesville native who assisted George Heye with logistics in western North Carolina, completely excavated the Notley Mound in Cherokee County (Turbyfill 1927). Turbyfill devotes a single paragraph to the excavation of the Notley Mound in a short paper on file at the National Museum of the American Indian (Turbyfill 1927).

In 1933 and 1934, the Smithsonian Institution, in conjunction with the Civil Works Administration, carried out extensive excavations at the Peachtree Mound near Murphy, North Carolina. The mound site was selected for its research potential but also because the area was in need of economic relief and had a temperate climate (Seltzer and Jennings 1941: 1). The Peachtree Mound was completely excavated, and Seltzer and Jennings (1941: 57) concluded that the site "is a component in which both Woodland and Mississippi traits occur simultaneously." Their report indicates that artifacts dating from approximately nine thousand years before the present to the eighteenth century were recovered at the site, but the chronology of the village and mound are still not very well understood.

The outbreak of World War II brought archaeological research to a halt, but beginning in the early 1960s the University of North Carolina began extensive surveys in western North Carolina in conjunction with the Cherokee Project, which was funded by the National Science Foundation in 1965. The goal of the Cherokee Project was to understand the development of Cherokee culture through a study of the archaeological record in western North Carolina (Dickens 1976; Keel 1976).

By the 1970s, these surveys and other fieldwork resulted in the documentation of over fifteen hundred archaeological sites in western North Carolina (Keel 1976; Ward and Davis 1999). These surveys provided a regional context for small-scale excavations (Keel 1976). The results of the Cherokee Project were published in theses, dissertations, books, and articles that became standard reference texts and created the framework for our current understanding of the archaeology of North Carolina (see Holden 1966; Egloff 1967; Dickens 1976; Keel 1976). These were important gains, but, ironically, there was very little formal involvement with Cherokee during the course of the Cherokee Project.

More recent research on mounds in western North Carolina has yielded new insight into individual sites and improved our understanding of the archaeology of the region. Archaeological survey and testing at the Nununyi, Birdtown, and Kituwah mounds on the Qualla Boundary indicate that the mound sites were occupied during the Mississippian period and the historic Cherokee period (Greene 1996, 1998; Riggs and Shumate 2003). These findings speak to the long-term indigenous occupation of the land that makes up the modern-day Cherokee reservation. Rodning's analyses of materials and records from the Coweeta Creek site have improved our understanding of Cherokee townhouses and domestic architecture (2009, 2010), and his analyses of pottery from Coweeta Creek have refined our definition of the Qualla ceramic series (2008).

Western Carolina University's ongoing research at the Spikebuck Mound and town site promises to shed new light on the ceramic chronology in the Hiwassee River drainage (Stout 2011). Research programs at the Biltmore Mound in Asheville (Kimball and Shumate 2003; Kimball, Whyte, and Crites 2010, 2013) and the Garden Creek site near Canton, North Carolina (Wright 2013) are generating new data for understanding western North Carolina's place in complex regional trade and exchange networks during the Middle Woodland period.

Despite these advances in mound research in western North Carolina, many basic research questions remain unanswered, especially at a regional scale. Prior to 2012, only sixteen mound sites were officially recorded with the state archaeological site file. By the 1980s, archaeologists working in western North Carolina had identified many of the better preserved mounds but had only carried out intensive research at a few sites and had made few systematic attempts to relocate damaged or destroyed mounds. Until the late 1990s, there were few attempts to involve the Cherokee community in archaeological research (Riggs 2002).

Fortunately, in recent years this picture has started to change. Starting in the late 1990s, and especially since the creation of the Tribal Historic Preservation Office of the Eastern Band of Cherokee Indians in 1999, there has been increased cooperation between archaeologists and the Cherokee community, particularly in the context of cultural resource management projects on the Qualla Boundary (Riggs 2002; Cooper 2009). Recent projects such as the archaeological survey of the Qualla Boundary (Greene 1996, 1998) and the survey and geophysical study of the Kituwah Mound (Riggs and Shumate 2003) are good examples.

The Eastern Band also funded and helped develop the research design for the Ravensford project, a large-scale data recovery project completed in advance of the construction of the new Cherokee K–12 school complex (Cooper 2009). In 1996 the Eastern Band purchased the Kituwah Mound, and in 2007 they acquired the Cowee Mound in partnership with the Little Tennessee Land Trust (Middleton 2011). In recent years, the University of Tennessee conducted a field school with Cherokee high school students at a multicomponent site in Great Smoky Mountains National Park (Angst 2012).

Not only have these successful collaborative efforts benefited the Eastern Band of Cherokee Indians, but they have also made a broader contribution to indigenous archaeology. A primary goal of indigenous archaeology is to encourage productive collaboration between archaeologists and indigenous communities (Colwell-Chanthaphonh et al. 2010; Croes 2010; Silliman 2010; Wilcox 2010). In practice these partnerships can take many forms. Indigenous and nonindigenous archaeologists and indigenous communities have worked together to develop research designs for archaeology projects (Colwell-Chanthaphonh et al. 2010; Curtis 2010; Townsend 2011). Indigenous craft experts have assisted nonindigenous archaeologists with artifact analysis (Thompson 2008; Croes 2010), generating more nuanced, accurate, and relevant interpretations of material culture. Collaboration and partnerships like these are resulting in archaeology that addresses the goals and needs and respects the traditional belief of indigenous communities while also producing more informed interpretations of the archaeological record.

Another important goal is to counter the "terminal narratives" (see Wilcox 2009) that depict Native Americans as vanished or conquered peoples. As the Native American archaeologist Michael Wilcox (2010: 224) writes:

> It is widely accepted that we either succumbed to massive epidemics, had been eliminated through warfare, or had "lost our culture" through missionization, acculturation, or forcible assimilation (Clifford 1990: 1–28). All change (referred to as "progress" in enlightened societies) is depicted as reductive or destructive in Indigenous societies. Any number of general textbooks on North American archaeology will list this tragic litany as the catastrophic fates of a marginalized people (Diamond 1996, 2005). The partition of prehistory and history as separate domains of study has only contributed to this imaginary rupture.

By relocating mounds and towns, which were important central places, and reconstructing indigenous settlement history in western North Carolina, we have the opportunity to empirically test the validity of the terminal narrative, rather than take it as given. We can reconstruct a clearer picture of when, how, and why people settled the Southern Appalachian landscape over the last two thousand years. Following the example of indigenous scholars like Wilcox (2010), as well as ethnohistorians and archaeologists who reconstructed Hernando de Soto's route through the southeastern interior (Hudson et al. 1985; Hudson 1997), this project helps break down the divide between history and prehistory in the Southern Appalachians. This will lead to a better understanding of Southern Appalachian settlement and counter unfounded a priori depictions of Cherokee as a conquered or marginalized people.

Initial Results

In addition to the historical problems of site destruction discussed above, a major barrier to understanding the prehistoric cultural landscape of western North Carolina is the lack of a concerted effort to compile all existing information about mound sites in a single location. This is a general problem in archaeological research and is hardly particular to western North Carolina. While the state site file contains an excellent database of archaeological sites and current site reports for the state, older records and finer scale data are harder to come by. Archival data and archaeological records and collections are scattered across universities, state offices, and museums, and possible mounds identified decades ago have not been revisited. Many historical references to Cherokee townhouses have not been cross-checked and confirmed with archaeological evidence. The fragmented nature of this information puts these important cultural resources at risk in the face of encroaching development.

The first step in this project was to examine all available archival sources for information about mounds and town sites in western North Carolina and compile this information into a single database containing accurate location data, archaeological and historical documentation, and preservation status for all the prehistoric and historic Cherokee mound and town sites in western North Carolina. This was completed in the summer of 2011, with the aid of archaeologists and historians from across the state.[1] The complete database and a summary report of these findings were filed with the Eastern Band of Cherokee Indians Tribal Historic Preservation Office and are now available to their staff as tools for their preservation, research, and outreach efforts.

The archival research suggested that while there were only sixteen known archaeological sites containing mounds or townhouses on file with the state, there may have been as many as sixty-eight mound and townhouse sites in the study area (Steere 2011, 2013). This finding contrasted with the prevailing notion that there were relatively few mound sites in the region and that fewer still could be identified archaeologically.

Following this archival research, archaeological fieldwork was carried out in the winter of 2011 and spring of 2012. Initial reconnaissance surveys were completed at thirty-seven locations to determine which of the newly identified possible mound sites were genuine (the remaining sites were inaccessible; most were on private property, and a few were inundated by lakes). The next stage of the project involved mapping and shovel test surveys to locate and describe unrecorded and poorly understood sites. In accordance with the research design developed with the Cherokee Tribal Historic Preservation Office, no invasive subsurface testing took place directly on known or possible mounds. Teams of archaeologists excavated shovel tests in the habitation areas around possible mounds to determine if sites were occupied by Cherokee and ancestral Cherokee people and to recover artifacts that could be used to assign dates of occupation to the sites. Existing ceramic collections and the new, systematic artifact collections were analyzed to assign approximate dates of occupation to sites.

During the reconnaissance survey, the principal investigator visited all of the possible mound sites, looking for evidence of mounds and villages, usually accompanied by local residents, archaeologists, and historians. It was determined that in addition to the sixteen previously recorded mound sites, thirty-four additional sites either contained mounds or were likely to have contained mounds that have been leveled.

From this group of sites, ten locations with known or possible Woodland, Mississippian, and/or Cherokee mounds were selected for intensive mapping and shovel testing with the goal of defining unknown or poorly understood site boundaries and generating ceramic samples for dating. This phase of the project resulted in the first modern mapping of several important mound sites. This includes the Notley or Notla Mound near Murphy, North Carolina, which was once over ten meters tall and may have been a major Early Mississippian period mound site (Turbyfill 1927), and the Whatoga or Watauga Mound, the remains of a Cherokee townhouse visited by the naturalist William Bartram in 1776 (Bartram 1928 [1775]). Our team also identified the location of the Jasper Allen Mound, a twelfth- or thirteenth-century mound site located near the modern town of Sylva, North Carolina, which was completely leveled by nineteenth-century excavations (Steere 2011, 2013, 2015).

The archaeological survey completed for this project revealed that eighteen of the sixty-eight archaeological sites identified through archival research lacked reliable archaeological or historical evidence for Woodland or Mississippian period mounds or Cherokee townhouses. Of the remaining fifty sites, twenty-five can be conclusively identified as containing Woodland or Mississippian period mounds or Cherokee townhouses. An additional twenty-five sites represent possible mound and/or townhouse locations, but further archival and archaeological research will be necessary to verify their status.

Discussion

The results of this project are best illustrated by the new maps of mound sites we have created (see Figure 8.3; for a more complete discussion of the results of this project, see Steere 2013 and 2015)

Given the preliminary nature of this research and the lack of fine-scale chronological information for the mound sites, few archaeological interpretations will be offered here, as they would be highly speculative. Rough dates of occupation have been assigned to only the twenty-five confirmed mounds, and in some cases these designations are tentative and, admittedly, based on limited data (e.g., reports from antiquarian excavations). However, a few key points merit discussion, even at this early stage of the project.

Our current understanding of Woodland period mound use in western North Carolina comes primarily from the long-term research at the Garden Creek site (Keel 1976; Wright 2013) and the Biltmore Mound (Kimball, Whyte, and Crites 2010, 2013). These studies revealed important information about Woodland period ceremonialism. However, western North Carolina's role in broader Woodland period social systems is still poorly understood.

Middle Woodland period mounds (ca. AD 200–800) are thought to be relatively uncommon in the Southern Appalachians, but this study suggests that there may have been more than ten Woodland period mounds or mound groups centered on the French Broad and Pigeon River drainages (see Figure 8.3). This spatial distribution of mound sites is similar to clusters of Woodland period ceremonial mounds and earthworks in the Midwest (Carr and Case 2006) and near the Leake site in northwest Georgia (Keith 2010). It is striking that there is only one possible Woodland period mound recorded west of the Pigeon River drainage (site 31GH35), and even this site lacks definitive archaeological evidence for a mound.

It is possible that Woodland period mounds, many of which do not exceed one or two meters in height and are easily destroyed by plowing and looting, may have existed in the western part of the study area and were destroyed before they could be recorded. However, it is also possi-

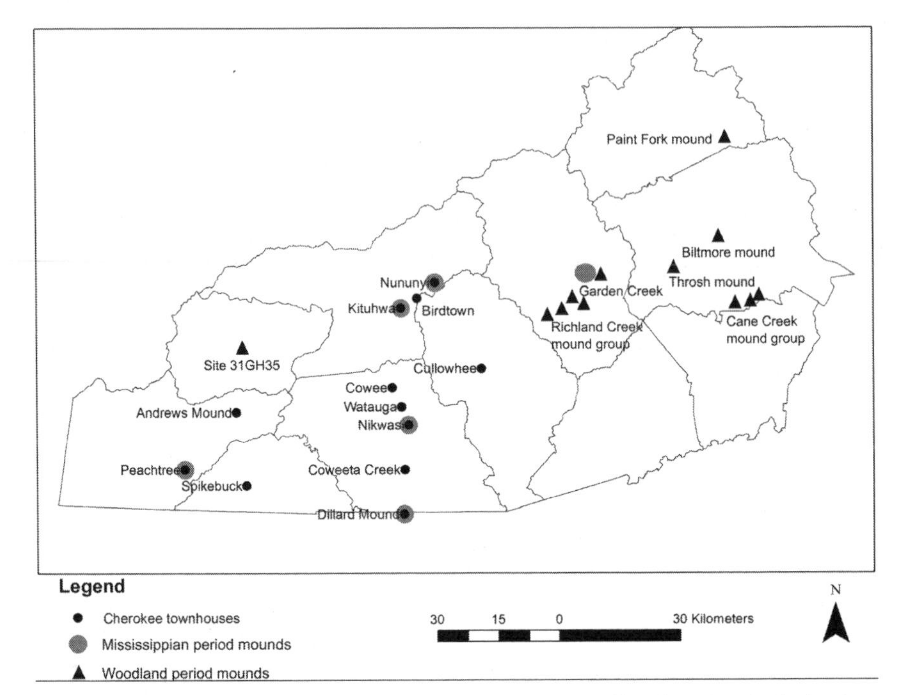

Figure 8.3. Map of Archaeologically Confirmed Woodland and Mississippian Mound Sites and Cherokee Townhouse Sites

ble that this distribution represents a genuine cultural pattern. If this is so, why was there significant Woodland period mound building activity in the Pigeon River and French Broad river valleys and apparently none in the Tuckasegee, Little Tennessee, and Valley River drainages? These sites may have been important nodes in the broad Hopewell Interaction Sphere (Keel 1976; Anderson and Mainfort 2002; Carr and Case 2006) and should be included in broader considerations of Woodland period life in the Southern Appalachian Mountains.

As Figure 8.3 shows, Mississippian period platform mounds were much more evenly distributed across the study area, with one or two mound sites each in the Pigeon, Tuckasegee, Little Tennessee, and Valley River drainages. This distribution of Mississippian period mounds is consistent with very general expectations for Mississippian period settlement patterns, especially David Hally's model for the territorial size of Mississippian polities (see B.D. Smith 1978, 1992; Anderson 1994; Hally 1996, 2006).

Hally (2006) used ceramic dating and mound stratigraphy to reconstruct the geography and timing of mound construction and occupation in northern Georgia during the Mississippian period. Based on the assumption that Mississippian platform mounds serve as proxies for the capitals of polities, Hally found that the minimum distance spacing separating neighboring, competing centers was thirty-five to fifty-five kilometers, that towns making up a chiefdom were generally situated along a river floodplain over a distance of ten to twenty kilometers, and that polities were separated by an unpopulated buffer zone measuring ten to thirty kilometers across (Hally 2006).

Following Hally's model (see Hally 1996, 2006), this would suggest that there may have been four Mississippian period polities in western North Carolina: one represented by the Pisgah phase mound at Garden Creek, a second by the Nununyi and Jasper Allen mounds in the Oconluftee and Tuckasegee drainages, a third by the Nikwasi Mound and the mound in Dillard, Georgia (see Elliot 2012) on the Little Tennessee River, and a fourth by the Peachtree and Notley mounds, located on tributaries of the Valley River.

The four clusters of mounds do not exceed forty kilometers in length, and they are separated by buffer zones of unoccupied territory, which in each case includes steep mountain ranges. At the moment, our understanding of the timing of the construction and use of these mounds is quite broad, but as the chronological associations for the known and possible mound sites are refined, it will be possible to test current models of Mississippian settlement patterns with new data from the region (see Steere 2015).

Figure 8.3 also shows the distribution of archaeologically identified Cherokee townhouses. In many cases, these locations were already known through previous historical and archaeological research (Smith 1979; Duncan and Riggs 2003; Gragson and Bolstad 2007; Boulware 2011). However, an important pattern emerges when the location of Mississippian period platform mounds and Cherokee townhouses are compared. Three of the four groups of Mississippian period platform mounds—the Peachtree and Notley mounds, the Nikwasi and Dillard mound, and the Nununyi and Jasper Allen bounds—appear to define the territories that would become the Valley, Middle, and Out Towns, respectively.

From a regional, long-term perspective, it appears that the eighteenth-century Cherokee towns are built "on top" of former Mississippian period polities, much in the same way that Cherokee townhouses are known to have been built on the summits of Mississippian period platform mounds at the Peachtree, Nikwasi, Dillard, and Nununyi mounds. The patterns observed here may provide regional scale archaeological support for the construction of Cherokee identity as a process of "emplacement," by which "a community attaches itself to a particular place through formal settlement plans, architecture, burials, and other material additions to the landscape" (Rodning 2009: 629). It seems likely that for historic period Cherokee communities, locations marked by Mississippian platform mounds were an especially important part of the cultural landscape and may have served as anchors for groups of towns.

Conclusions

These preliminary findings have already shed new light on the nature of sacred Cherokee sites in western North Carolina. At the most basic level, the new location data for mounds once considered lost in the archaeological record allow us to relocate these important sites, places that contain the burials of ancestral Cherokee and objects such as sacred central hearths. More specifically, the new findings paint a portrait of a more complex and expansive built environment in the Cherokee heartland than those depicted in earlier archaeological reconstructions for the region. Additional research that builds on this initial study will no doubt improve our understanding of change and continuity in the nature of the sacred landscape in the Cherokee heartland over nearly two millennia.

Beyond generating new information for archaeological research and preservation, this project can serve as a model for positive collaborative research between archaeologists and indigenous communities in the service of native interests (see Riggs 2002). A major critique of archaeological research, and one that still applies today even after the passage of the Native American Graves Protection and Repatriation Act of 1990, is that archaeology is something done to, not with, or for, indigenous groups (Watkins 2000). This project is designed to use the tools of archaeology to give something back to the Cherokee community.

First, the database from this project will serve as a monitoring tool for the Tribal Historic Preservation Office of the Eastern Band of Cherokee Indians. Mound and town sites, even those that have been badly disturbed, have a high probability of containing graves. With updated status and location information for these sites, many of which are currently lost, the staff of the Tribal Historic Preservation Office will be better equipped to carry out their stewardship responsibility.

Second, this project will help prioritize sites for preservation and land acquisition and will build on the success of projects like the acquisition of the Kituwah and Cowee mounds. Sacred sites in western North Carolina will continue to be threatened by development and other destructive processes in the coming years. Knowing the location and current preservation status of mound and town sites will enable the Eastern Band to make more informed decisions about which sites to purchase and preserve.

Finally, the results of this project will be used to expand our understanding of the historical geography of the Cherokee landscape. This new knowledge can be used to protect and enhance Cherokee cultural identity, and it makes a significant contribution to the development of indigenous archaeology for the Eastern Band of Cherokee Indians. Most members of the Cherokee community are intimately familiar with the location and meaning of important sites on or near the Qualla Boundary, such as the Kituwah Mound. However, members of the Eastern Band are probably not as familiar with the names, locations, and stories of important mound and town sites outside the Qualla Boundary, such as the mounds along the Little Tennessee or Valley rivers that were leveled by late nineteenth-century antiquarian expeditions. This narrower view of Cherokee historical geography is at once the result of the violence and land cessions of centuries past and the more recent destruction of important places.

We have already begun to use the information generated by this project for public outreach efforts in the Cherokee community and beyond. The staff of the Tribal Historic Preservation Office and I have presented the results of our work at community club meetings in Cherokee and other public venues such as libraries and community centers in neighboring counties. In addition to presenting the findings of this project in scholarly journals and professional conferences, we shared the initial results of our work with members of the Eastern Band at the second and third annual Cherokee Archaeology Day symposia in 2012 and 2013. These public archaeology events, organized by the Tribal Historic Preservation Office and held annually on the Qualla Boundary, provide an opportunity for archaeologists working in western North Carolina to share their work with members of the Cherokee community.

By relocating and studying these places archaeologically, we can create a broader reconstruction of the Cherokee world before contact and removal. Mounds are a physical connection to Cherokee cultural identity, material reminders of past and present Cherokee lifeways and traditions. Some mounds and townhouses that were damaged by plowing, development, and antiquarian explorations may still be partially intact and are still important, living places on the landscape. Putting these places back on the map is an important step for revitalizing Cherokee culture.

Acknowledgments

The author would like to thank Fausto Sarmiento and Sarah Hitchner for organizing the sacred sites conference and this edited volume. It was a privilege and pleasure to be invited to contribute to this effort. I would also like to thank John Chamblee and Ted Gragson at the Coweeta Long Term Ecological Research Program at the University of Georgia for their support of this research. This project was developed in close cooperation with the Tribal Historic Preservation Office of the Eastern Band of Cherokee Indians, and I would like to thank Russell Townsend, Brian Burgess, Tyler Howe, Yolanda Saunooke, Miranda Panther, Beau Carroll, and Johi Griffin for their role in designing and supporting the research, public outreach, and preservation efforts of the Western North Carolina Mounds and Towns Project. Russell Townsend's support and guidance were especially important, and I so appreciate his help.

This research was supported by a National Science Foundation award (DEB-0823293) from the Long Term Ecological Research Program to the Coweeta LTER Program at the University of Georgia. Any opinions, findings, conclusions, or recommendations expressed in the material are those of the author and do not necessarily reflect the views of the National Science Foundation or the University of Georgia. This research was also supported by the Cherokee Preservation Foundation, the Tribal Historic Preservation Office of the Eastern Band of Cherokee Indians, the Duke Energy Foundation, TRC Environmental Corporation, the University of West Georgia, and Western Carolina University. Many people helped improve this article; any errors herein are the responsibility of the author.

Benjamin A. Steere, Ph.D., is an assistant professor of anthropology at Western Carolina University, Cullowhee, North Carolina, and an expert on applied ethnographic research and archaeology in the Appalachians.

Notes

1. Between 21 June and 31 August 2011, the project director traveled to the following locations to carry out archival research and interviews: the Tribal Historic Preservation Office in Cherokee; the North Carolina Office of State Archaeology (OSA) in Raleigh and the western branch of the OSA in Asheville; the North Carolina State Archives in Raleigh; the Research Laboratories of Archaeology at University of North Carolina Chapel Hill (UNC RLA); Western Carolina University in Cullowhee; the archives of the National Park Service at the Great Smoky Mountains National Park visitor center in Gatlinburg, Tennessee; the Franklin Press in Franklin; the North Carolina Rooms of the Buncombe, Henderson and Haywood County libraries; the office of the register of deeds in Buncombe, Henderson, and Jackson County and the Main Library; and the Map Library at the University of Georgia. The results from the archival research for this project are discussed in detail in reports of pre-

liminary research for the Western North Carolina Mound and Towns Project (Steere 2011, 2013).

Background research was also conducted using geographic information systems (GIS) available publically through county land record websites and other sources, such as the North Carolina Department of Transportation (NCDOT) website and the North Carolina State University GIS clearinghouse. LIDAR data available through the NCDOT website was especially useful for identifying and assessing possible mound locations.

References

Anderson, David G. 1994. *The Savannah River Chiefdoms: Political Change in the Late Prehistoric Southeast.* Tuscaloosa: University of Alabama Press.

Anderson, David G. and Robert C. Mainfort, Jr. 2002. "An Introduction to Woodland Archaeology in the Southeast." In *The Woodland Southeast,* edited by David G. Anderson and Robert C. Mainfort, Jr., 1–19. Tuscaloosa: University of Alabama Press.

Angst, Michael G. 2012. "Archaeological Investigations at Site 31SW393, Smokemont, Swain County, North Carolina." Unpublished report prepared for Great Smoky Mountains National Park, Gatlinburg, Tennessee.

Bartram, William. 1928 [1775]. *Travels of William Bartram.* New York: Dover.

Beck, Robin A., Jr. 2003. "Consolidation and Hierarchy: Chiefdom Variability in the Mississippian Southeast." *American Antiquity* 68, no. 4: 641–661.

Booker, Karen M., Charles M. Hudson, and Robert L. Rankin. 1992. "Place Name Identification and Multilingualism in the Sixteenth-Century Southeast." *Ethnohistory* 39, no. 4: 399–451.

Boulware, Tyler. 2011. *Deconstructing the Cherokee Nation: Town, Region, and Nation among Eighteenth-Century Cherokees.* Gainesville: University Press of Florida.

Carr, Christopher and D. Troy Case, eds. 2006. *Gathering Hopewell: Society, Ritual, and Ritual Interaction.* New York: Springer.

Clifford, James. 1990. *The Invented Indian: Cultural Fictions and Governmental Policies.* New Brunswick, NJ: Transaction Press.

Cooper, Andrea. 2009. "Embracing Archaeology." *American Archaeology* 13, no. 3: 19–24.

Colwell-Chanthaphonh, Chip, T.J. Ferguson, Dorothy Lippert, Randall H. McGuire, George P. Nicholas, Joe E. Watkins, and Larry J. Zimmerman. 2010. "The Premise and Promise of Indigenous Archaeology." *American Antiquity* 75, no. 2: 228–238.

Croes, Dale R. 2010. "Courage and Thoughtful Scholarship = Indigenous Archaeology Partnerships." *American Antiquity* 75, no. 2: 211–216.

Curtis, Wayne. 2010. "The Development of Indigenous Archaeology." *American Archaeology* 14, no. 3: 37–43.

Diamond, Jared. 1996. *Guns, Germs, and Steel: The Fates of Human Societies.* New York: Norton.

———. 2005. *Collapse: How Societies Choose to Fail or Succeed.* New York: Viking.

Diaz-Granados, Carol. 2004. "Marking Stone, Land, Body, and Spirit: Rock Art and Mississippian Iconography." In *Hero, Hawk, and Open Hand: American Indian Art of the Ancient Midwest and South,* edited by Richard F. Townsend, 139–149. New Haven: Yale University Press and the Art Institute of Chicago.

Dickens, Roy S., Jr. 1976. *Cherokee Prehistory: The Pisgah Phase in the Appalachian Summit Region.* Knoxville: University of Tennessee Press.

Duncan, Barbara R. and Brett H. Riggs. 2003. *Cherokee Heritage Trails Guidebook.* Chapel Hill: University of North Carolina Press.

Egloff, Brian J. 1967. "An Analysis of Ceramics from Historic Cherokee Towns." Unpublished Master's Thesis. Department of Anthropology, University of North Carolina, Chapel Hill.

Elliot, Daniel T. 2012. *Colburn's Excavation of the Greenwood Mound (a.k.a. J. J. Greenwood Mound), Rabun County, Georgia.* LAMAR Institute Research Publication 115. Athens, GA: LAMAR Institute.

Fenneman, Nevin M. 1938. *Physiography of Eastern United States.* New York: McGraw-Hill.

Finger, John R. 1984. *The Eastern Band of Cherokee, 1891–1900.* Knoxville: University of Tennessee Press.

Gragson, Ted L. and Paul V. Bolstad. 2007. "A Local Analysis of Early Eighteenth-Century Cherokee Settlement." *Social Science History* 31, no. 3: 435–468.

Greene, Lance K. 1996. "The Archaeology and History of the Cherokee Out Towns." Unpublished Master's Thesis. Department of Anthropology, University of Tennessee, Knoxville.

———. 1998. "An Archaeological Survey of the Qualla Boundary in Swain and Jackson Counties, North Carolina." Unpublished report submitted to the Office of State Archaeology, Raleigh.

Hally, David J. 1996. "Platform-Mound Construction and the Instability of Mississippian Chiefdoms." In *Political Structure and Change in the Prehistoric Southeastern United States,* edited by J.F. Scarry, 92–127. Gainesville: University of Florida Press.

———. 2006. "The Nature of Mississippian Regional Systems." In *Light on the Path: The Anthropology and History of the Southeastern Indians,* edited by T.J. Pluckhahn and R. Ethridge, 26–42. Tuscaloosa: University of Alabama Press.

Hally, David J. and Robert C. Mainfort. 2004. "Prehistory of the Eastern Interior after 500 B.C." In *Southeast,* edited by R.D. Fogelson, 265–285. Handbook of North American Indians, vol. 14. Washington: Smithsonian Institution.

Heye, George C. 1919. "Certain Mounds in Haywood County, North Carolina." *Contributions from the Museum of the American Indian, Heye Foundation* 5, no. 3: 35–43.

Holden, Patricia P. 1966. "An Archaeological Survey of Transylvania County, North Carolina." Unpublished Master's Thesis. Department of Anthropology, University of North Carolina, Chapel Hill.

Hudson, Charles M. 1997. *Knights of Spain, Warriors of the Sun: Hernando de Soto and the South's Ancient Chiefdoms.* Athens: University of Georgia Press.

Hudson, Charles M., Marvin T. Smith, David J. Hally, Richard R. Polhemus, and Chester B. DePratter. 1985. "Coosa: A Chiefdom in the Sixteenth-Century Southeastern United States." *American Antiquity* 50, no. 4: 723–737.

Keel, Bennie C. 1976. *Cherokee Archaeology: A Study of the Appalachian Summit.* Knoxville: University of Tennessee Press.

Keith, Scot J. 2010. "Archaeological Data Recovery at the Leake Site, Bartow County, Georgia." Unpublished report prepared for the Georgia Department of Transportation Office of Environment/Location, Atlanta.

Kimball, Larry R. and Scott M. Shumate. 2003. "2002 Archaeological Excavations at the Biltmore Mound, Biltmore Estate, North Carolina." Unpublished report submitted to the Committee for Research and Exploration, National Geographic Society, Washington, D.C.

Kimball, Larry R., Thomas R. Whyte, and Gary D. Crites. 2010. "The Biltmore Mound and Hopewellian Mound Use in the Southern Appalachians." *Southeastern Archaeology* 29, no. 1: 44–58.

Kimball, Larry R., Thomas R. Whyte, and Gary D. Crites. 2013. "Biltmore Mound and the Appalachian Summit Hopewell." In *Early and Middle Woodland Landscapes of the Southeast,* edited by Alice P. Wright and Edward R. Henry, 122–137. Gainesville: University of Florida Press.

King, Adam. 2003. *Etowah: The Political History of a Chiefdom Capital.* Tuscaloosa: Alabama Press.

King, Duane H., editor. 2007. *The Memoirs of Lt. Henry Timberlake: The Story of a Soldier, Adventurer, and Emissary to the Cherokees, 1756-1765.* Cherokee: Museum of the Cherokee Indian Press.

Knight, Vernon James, Jr. 2010. *Mound Excavations at Moundville: Architecture, Elites and Social Order.* Tuscaloosa: Alabama Press.

Lindauer, O. and John H. Blitz. 1997. "Higher Ground: The Archaeology of North American Platform Mounds." *Journal of Archaeological Research* 5, no. 2: 169–207.

Middleton, Beth Rose. 2011. *Trust in the Land: New Directions in Tribal Conservation.* Tucson: University of Arizona Press.

Milner, George. 2004. *The Moundbuilders: Ancient Peoples of Eastern North America.* London: Thames & Hudson.

Mooney, James. 1900. *Myths of the Cherokee. Nineteenth Annual Report of the Bureau of American Ethnology, 1897–1898, Pt. 1.* Washington, D.C.: Smithsonian Institution.

Muller, Jon. 1997. *Mississippian Political Economy.* New York: Plenum Press.

Perdue, Theda. 1998. *Cherokee Women: Gender and Culture Change, 1700–1835.* Lincoln: University of Nebraska Press.

Perdue, Theda and Michael D. Green. 2001. *The Columbia Guide to American Indians of the Southeast.* New York: Columbian University Press.

Putnam, F.W. 1884. "Seventeenth Report." *Reports of the Peabody Museum of American Archaeology and Ethnology* 3: 339–384.

Riggs, Brett. 2002. "In the Service of Native Interests: Archaeology for, of, and by Cherokee People." In *Southern Indians and Anthropologists: Culture, Politics, and Identity,* edited by Lisa J. Lefler and Frederic W. Gleach, 19–30. Athens: University of Georgia Press.

Riggs, Brett H. and M. Scott Shumate. 2003. *Archaeological Testing at Kituwah. 2001 Investigations at Sites 31Sw1, 31Sw2, 31Sw287, 31Sw316, 31Sw317, 31Sw318, and 31Sw320.* Unpublished report prepared for the Eastern Band of Cherokee Indians Cultural Resources Program. Report on file, NC Office of State Archaeology.
Rodning, Christopher B. 2002. "The Townhouse at Coweeta Creek." *Southeastern Archaeology* 21, no. 1: 10–20.
———. 2008. "Temporal Variation in Qualla Pottery at Coweeta Creek." *North Carolina Archaeology* 57: 1–49.
———. 2009. "Mounds, Myths, and Cherokee Townhouses in Southwestern North Carolina." *American Antiquity* 74, no. 4: 627–663.
———. 2010. "Architectural Symbolism and Cherokee Townhouses." *Southeastern Archaeology* 29, no. 1: 59–80.
———. 2015. *Center Places and Cherokee Towns: Archaeological Perspectives on Native American Architecture and Landscape in the Southern Appalachians.* Tuscaloosa: University of Alabama Press.
Setzler, Frank M. and Jesse D. Jennings. 1941. *Peachtree Mound and Village Site, Cherokee County, North Carolina.* Bureau of American Ethnology Bulletin 131. Washington, D.C.: Smithsonian Institution.
Silliman, Stephen W. 2010. "The Value and Diversity of Indigenous Archaeology: A Response to McGhee." *American Antiquity* 75, no. 2: 17–220.
Smith, Betty Anderson. 1979. "Distribution of Eighteenth-Century Cherokee Settlements." In *The Cherokee Indian Nation: A Troubled History,* edited by Duane H. King, 46–60. Knoxville: University of Tennessee Press.
Smith, Bruce D., ed. 1978. *Mississippian Settlement Patterns.* New York: Academic Press.
———. 1992. *Rivers of Change: Essays on Early Agriculture in Eastern North America.* Washington, D.C.: Smithsonian Institution Press.
Steere, Benjamin A. 2011. "Preliminary Results of Archival Research for the Western North Carolina Mounds and Towns Project: Documentary Evidence for Known and Potential Mound Sites in Buncombe, Cherokee, Clay, Graham, Haywood, Henderson, Jackson, Macon, Madison, Swain, and Transylvania Counties, North Carolina." Report on file, Eastern Band of Cherokee Indians Tribal Historic Preservation Office.
———. 2013. "The Western North Carolina Mounds and Towns Project: Results of 2011–2012 Archival Research and Field Investigations in Buncombe, Cherokee, Clay, Graham, Haywood, Henderson, Jackson, Macon, Madison, Swain, and Transylvania Counties, North Carolina." Report on file, Eastern Band of Cherokee Indians Tribal Historic Preservation Office.
———. 2015. "Revisiting Platform Mounds and Townhouses in the Cherokee Heartland: A Collaborative Approach." *Southeastern Archaeology* 34, no. 3: 196–219.
Steere, Benjamin A., Paul A. Webb, and Bruce S. Idol. 2012. "A 'New' Account of Mound and Village Sites in Western North Carolina: The Travels of Captain R.D. Wainwright." *North Carolina Archaeology* 61: 1–37.
Stout, Andy. 2011. "A Piece of Cherokee History: The Conservancy Signs an Option for a Significant Cherokee Town Site." *American Archaeology* 15 (spring): 44–45.

Thomas, Cyrus. 1891. *Catalogue of Prehistoric Works East of the Rocky Mountains.* Bureau of American Ethnology Bulletin 12. Washington, D.C.: Smithsonian Institution.

———. 1894. *Reports on the Mound Explorations of the Bureau of American Ethnology.* Twelfth Annual Report of the Bureau of Ethnology, 1890–1891. Washington, D.C.: Smithsonian Institution.

Thompson, Ian. 2008. *Chahta Intikba Aiikhvna Learning from the Choctaw Ancestors: Integrating Indigenous and Experimental Approaches in the Study of Mississippian Technology.* Ph.D. Dissertation. University of New Mexico.

Townsend, Russell. 2011. "Native Soil: A Cherokee Archaeologist Digs into His Own Heritage." *SAA Archaeological Record* 11 (March): 21–23.

Turbyfill, Charles O. 1927. "Work Done in Western North Carolina during the Summer of 1926." Manuscript on file at the Archives of the National Museum of the American Indian, Washington, D.C.

Valentine, G.G., B.B. Valentine, and E.P. Valentine. 1889. "Catalogue of Objects. In The Valentine Museum" [Museum handbook]. Richmond, VA. On file at the Research Laboratories of Archaeology, University of North Carolina, Chapel Hill.

Wainwright, Robert D. 1913. "A Summer's Archaeological Research." *Archaeological Bulletin* 4, no. 5: 111–121.

———. 1914a. "A Summer's Archaeological Research." *Archaeological Bulletin* 5, no. 1: 6–9.

———. 1914b. "A Summer's Archaeological Research." *Archaeological Bulletin* 5, no. 2: 29–30.

Ward, H. Traywick and R.P. Stephen Davis. 1999. *Time before History: The Archaeology of North Carolina.* Chapel Hill: University of North Carolina Press.

Watkins, Joe. 2000. *Indigenous Archaeology: American Indian Values and Scientific Practice.* Walnut Creek: Altamira Press.

Wilcox, Michael. 2009. *The Pueblo Revolt and the Mythology of Conquest: An Indigenous Archaeology of Contact.* Berkeley: University of California Press.

———. 2010. "Saving Indigenous Peoples from Ourselves: Separate but Equal Archaeology Is Not Scientific Archaeology." *American Antiquity* 75, no. 2: 221–227.

Wright, Alice P. 2013. "Under the Mound: The Early Life History of the Garden Creek Mound No. 2 Site." *In Early and Middle Woodland Landscapes of the Southeast,* edited by Alice P. Wright and Edward R. Henry, 108–121. Gainesville: University Press of Florida.

———. 2014. "History, Monumentality, and Interaction in the Appalachian Summit Middle Woodland." *American Antiquity* 79, no. 2: 277–294.

CHAPTER 9

Biocultural Sacred Sites in Mexico

Mindahi Crescencio Bastida Muñoz
and Geraldine Patrick Encina

Introduction

There have been many different approaches and contributions to our intercultural understanding of the inherent sacredness of nature. In attempting to strengthen the rationale for what constitutes a sacred natural site (SNS), Mercedes Otegui-Acha (2007) provided a series of variables that characterize SNS. The variables she considered were (1) the intrinsic capacity of SNS to sustain biodiversity, (2) the quality of SNS to support indigenous and traditional peoples' ways of life (by keeping alive heritage, cultural identity, ethnolinguistic diversity, livelihoods, and traditional ecological knowledge), (3) the support of indigenous and traditional peoples' spirituality provided by SNS, and (4) the policy contexts that enable conservation of SNS.

After a thorough revision, Otegui-Acha (2007) presented a framework for developing an inventory of SNS. To determine their geographical distribution, she first produced a classification of SNS according to four spatial scales: (1) spatially dispersed sacred landscape, (2) spatially definable sacred landscape, (3) sacred natural physiographical features, and (4) sacred floral and faunal species. Together with a team of researchers from PRONATURA (an environmental nongovernmental organization based in Mexico), she ventured to test the methodology and tools, leading to the eventual creation of the National Inventory of Sacred Natural Sites in Mexico in 2010 (Otegui-Acha et al. 2010).

We propose that the framework used by Otegui-Acha (2007) is missing three concepts that are crucial in Mesoamerican cosmo-vision:[1] (1) bioculture, (2) time–space, and (3) energetic activation. These three concepts are so inherent in native peoples' lives that their interlocutors do not feel the need to delineate them in abstract forms. However, in an intercultural arena, where different parties are trying to construct an equitable approach to conservation of these sites, such concepts need to be brought forward. The ideal scenario is that, as practitioners of transdisciplinary research, we can directly and indirectly engage in the native peoples' ways of being (of living and interrelating), allowing us all to forge, in synergic and sensible ways, the safeguarding of culturally relevant places.

First Concept: Bioculture

Human thought about existence and transcendence does not flow and expand itself regardless of experience in nature, nor by "taking oneself out of the world" to construct a view of the world, as Ingold (2000: 95) states. This axiom is at least true among original (or indigenous) peoples[2] and their descendants. Drawing from constructivism and human geography, we note that any natural element on earth or in the sky that is apprehended by the human heart and mind becomes cultural. Going a step further, collective members from original cultures perceive the world as all living (Lenkersdorf 1999). For instance, Mapuche peoples from Chile and Argentina call "beings" (*che*) all those who interplay as agents or simply as related entities in the cycle of life: rocks, fungi, mosses, and plants and animals of water, land, and sky.

In Maya territory, the Ceiba is not merely and botanically "a tree": it is the holder of one of the corners of the world. It is a being who has the willingness to do the job and who gets tired and lets go of the weight every once in a while. We propose that every element from nature, be it "tangible" or "intangible," "biological," "mineral," "from underground," "earthly," or "heavenly,"[3] is a biocultural element,[4] emphasizing the fact that original peoples engage with the world as a living entity. Ingold's (2000) explanation about relational thinking applies not only for subsistence hunters; among Pueblo Indians, "all my relations" is a common salutation, which reflects their particular way of experiencing their existence.

The kind of culture that relates to the natural world by seeing vital energy flowing through all its beings is a "bioculture." Tangible and intangible aspects of plants, fungi, animals, stones, rocks, water springs, flowing or still waters, mountains, caves, and landscapes, as experienced by original peoples in a life cycle context, become their biocultural heritage. We are emphasizing "their" because it is what they love and value with their eyes, their minds, and their hearts. It is very difficult for nonbiocultures to apprehend what biocultural heritage actually entails.

The term "biocultural heritage" has been used lately by ethnoecologists such as Toledo, Boege, and Bassols (2010) to refer to all that native peoples have produced, reproduced, and preserved in their territories as a result of the beliefs, knowledge, and practices emerging from their cultures. As used here, the concept emphasizes the idea that culture has developed around biological components and whatever nurtures life: soil, land, wind, and water (Barrera-Bassols 2003).

Toledo (2002) proposes that the diversity of animals and plants (biodiversity), of spoken languages (ethnodiversity), and of regions of domesticated plants and animals (agrodiversity) together represent a biocultural heritage. He stresses the interconnections among those dimensions, since it is in their blending that a complex system merges and evolves. These interconnections result in (1) the geographic overlap between biological and linguistic diversities; (2) the overlapping between native peoples' territories and biologically rich regions; (3) native peoples as the main inhabitants and managers of well-preserved landscapes; and (4) the certification of a behavior oriented to the conservative use of landscapes among original peoples derived from their beliefs, knowledge, and practices (Toledo et al. 2001).

When these aspects are taken into account, Mexico is the country with the second highest number of biocultural sites in the world (Toledo 2012). The geographical spots of the main cultures of Mexico are found where biologically diverse habitats exist, and they are known as "biocultural regions" (Toledo 1999, 2007). In effect, we can find different peoples inhabiting a plethora of ecosystems and being so united with their landscape that they are called coast cultures, mountain cultures, jungle cultures, desert cultures, tropical wetland cultures, lake cultures, and highlands cultures. Most sacred places in Mexico are found in biocultural regions. Further below we propose which elements need to converge for a place to be sacred.

The category of "biocultural region" is slowly being used more among scholars. For over a decade, it has become fairly common to read about mindscapes or "langscapes" from Barrera-Bassols (2003) or Maffi (2001), but we still don't read about biocultural regions. This geographical category is quite natural, if seen from an integrated perspective. The superspecialization of sciences and of derived governmental programs has led to serious problems that end up in what we term "ecoethnocides."

Programs promoted by international initiatives, ministries, or other institutions often require experts or consultants to identify either "natural sites" or "cultural sites," and in Mexico the INAH (National Institute of Anthropology and History) has a different agenda from that of the SEMARNAT (Secretariat of Environment and Natural Resources). These independent efforts represent a dichotomy that originated from splitting a single epistemic approach into two separate fields of knowledge, namely, ecology and anthropology. Of deeper concern is the fact that neither of those fields is able to apprehend the biocultural world. Both can, with their conceptual frameworks, define a site as naturally or culturally important, but that definition does not prevent sites from becoming more and more eroded.

There must be a transdisciplinary approach that includes original people who can share their worldview and help others comprehend the implications of stepping into the new paradigm that is being built by human ecologists like Kassam (2009), ethnoecologists like Toledo and Barrera-Bassols (2008), critical sociologists like Boaventura de Sousa Santos (2010), and new ecologists like Berkes, Colding, and Folke (2003). A paradigm that is built on native conceptions of nature as a living entity is certainly and urgently needed in the minds of those who are ready to help slow down, stop, and reverse the deterioration of biocultural territories.

A biocultural heritage is not merely the happy result of original peoples' ways of life, which nation–states may feel is there for them to proudly display in world statistics or photographs. What has been passed on from original ancestors is a very ancient way of life that sees life flowing through the tangible and intangible, in a biocultural world. That way of life is continually experienced by biocultures. Those material and nonmaterial elements—all of which can be named and contacted through the mind and heart—are biocultural elements that collectively represent a biocultural heritage, one that primarily makes sense to the bioculture that signifies it.

Interpretations of such heritage will always have a margin of error, which interpreters must humbly accept. In this sense, they will have to provide conditions for original people to speak for themselves about what a territory, however large or small, however impressive or apparently insignificant, means to them. They must be given respect and the power to avoid the advancement of any kind of exploitation, extraction, or construction project. At the very least, the principle of free prior informed consent (FPIC) must be applied when projects are being conceived or are underway.

Second Concept: Time–Space

Any particular place is signified by a particular time, and any one time is signified by a specific place. Units of time–space are orderly, set in a single lattice, and this lattice helps contextualize a bioculture. Thus, the place that gives identity to a bioculture is a biocultural site. The term "site" speaks not only of a geographic coordinate but also of a meeting point of time and space.

This chronotopic axiom is absolutely crucial if we are to produce an inventory that is culturally meaningful in Mesoamerica. More importantly, it makes sense to native elders and also to youth who want to recover the ancestral conception of time–space.[5] They need to get involved in the process of understanding ancient astronomic calculations and practices because such calculations led to the erection of buildings or the selection of particular natural mounds or caves. Sites that are annually activated by means of rituals can help maintain the knowledge system of how celestial bodies determine meteorological cycles if there are people who actively study the time–space logic behind those sites.

This knowledge used to be traditionally reproduced, but ever since the arrival of Spanish friars, any means of transmission of this knowledge was severely punished. Today, common people participating in rituals only focus on deities with syncretic meanings, and they are not conscious of the astronomic and calendric implications of the ritual or of its symbolic role as a means to control weather and life cycles. In spite of that tendency, we must highlight that there do exist specialized groups of timekeepers who are knowledgeable in these matters.

From the Preclassic period to the late Postclassic period in Mesoamerica, wherever rituals were carried out, there was a permanent interplay between religious and astronomical motivations; like time and space, science and religion were fused, and together they explained the whole world. A distinct feature of prevailing Mesoamerican societies today is the intertwining of the explicit and the implicit; the tangible and intangible are conjoined in their world conception. But a major disruption occurred over the past centuries, when science and religion were split to explain the world. Even more, it was decreed that science was only possible among Western thinkers, so the rest of human thought was considered "pseudoscience," and it was stressed that religion was only for those institutions with authority to contact the divine. In this modern context, Mesoamerican calendars became merely ritual or divinatory and were considered almanacs that did not necessarily have to be backed up by a scientific explanation.

However, time–space knowledge and all the methodology around it (including high-precision instruments such as the alignment of bodies on the landscape and the development of acute observational techniques around the movement of celestial bodies) do have astronomic and meteorological implications, not only in Mesoamerica but also in Arid America (as described in what follows) and beyond. Astronomic referents (the Milky Way, the Pleiades, Venus, the Moon, and the Sun) are used as meteorological marking points and must be observed from specific sites.

For instance, Enrique Aguilar[6] reports that the site of Wirikuta—where the Jicuri (peyote, *Lophophora williamsii*) is harvested—is highly significant because the ceremonial act of sticking an arrow in the core of a Jicuri flower symbolizes the killing of a deer, which in turn symbolizes killing the drought. So putting an end to the deer peyote on this earthly level puts an end to the dry aspect of the heavens (and by fractal law, to the dry aspect of the sky level). When the peyote is brought back from Wirikuta, the Jicuri Neixa celebration takes place in order to present the blue flag, which represents the rain that has been brought from the East.

Through the various rituals along the pilgrimage route, time–space is placed in order, and only then is everything set in place for the rainy season to commence: the date is 12 to 13 May, twenty days before the solstice on 21 June (Enrique Aguilar 2012, personal communication). What has happened in the heavens or on the cosmic level is that the Milky Way, announced by the Pleiades on the East at dawn, has started to become visible. Over the last few millennia, the thick trail of water drops (the Milky Way as conceived by Hñahñu-Otomí peoples[7]) has helped keep track of the return of the wet season. How much rain will fall in a certain year depends on how blurry the Milky Way and the Pleiades appear. This is just one example of hundreds that sustain the referred to chronotopic couple.

The approach necessarily requires the respectful knowledge and handling of the Mesoamerican chronotopic instrument: the "eighteen twenty-day month plus five days" calendar (i.e., the year calendar), together with the "thirteen-times-twenty-days" cycle (the 260-day cycle). We are only just now able to fully grasp the time–space conception thanks to Patrick's recent findings (Patrick 2013): in short, the 365-day cycle is always fixed to the yearly Sun cycle because each Mesoamerican day is a complete day. This disputes the thesis stating that the Mesoamerican year is vague, wandering forever around the solar cycle. Moreover, the 260-day moving cycle provides the same name to any particular day all throughout Mesoamerica. This 260-day cycle is one of a kind: a robust lattice conveyed and perpetuated by a singly rooted myriad of cultures that, by providing concerted rituals, helped to keep the lattice in place for at least three millennia.

The fundamental structure of conceived time–space is supported by four trees at each corner and one in the middle. The middle one is physically represented by the hearth, where the eternal fire is lit and relit each year (on 22 March). Each of the four trees symbolizes a position of the Sun both in its daily course—sunrise, midday, sunset, and midnight—and in its yearly course. This yearly cycle produces spatial referents (eastern segment, northern segment, western segment, southern segment[8]) and also helps to identify dates concerning significant moments in the weather and in corn growth. Thus, a multifactor correspondence can be made, whereby:

1. 10 to 12 February is linked to sunrise, to the eastern portion of the landscape, to the blessing of corn seeds (and also to the end and beginning of the Mexica year, a culture focusing on the rising Sun on the East).
2. 30 April to 3 May is linked to midday, to the northern portion of the sky and the land, to the beginning of the rainy season (and to the second zenith passage over the southernmost area of Mesoamerica, which happens 260 days after the first zenith passage, on 13 August).
3. 13 August is linked to sunset, to the western segment of the territory, to the interfestival drought within the rainy season (and to the beginning of the Maya year).
4. 29 October to 2 November is linked to midnight, to the southern segment of the world, to the underworld, to the end of the rainy season (and to the end of the corn plant lifecycle, which ritually began on 12 February).

As demonstrated in Figure 9.1, there are two coupled sets of dates that are 260 days apart. One couple is 10 to 12 February through 29 October to 2 November, and it was originally Pleaides based. It is a cycle evoked in many Central Mexican stelae by means of two calendar glyphs: 1 Rabbit and 2 Reed. They mean that 260 days are counted from day 1 Rabbit of year 1 Rabbit (i.e., from the day acting as the year bearer[9] of the year to be completed, which falls on 10/11 February[10]) until day 1 Rabbit of year 2 Reed, which falls on 28/29 October. This is the interval necessary to view the journey of Pleiades from being directly overhead at sunset to being directly overhead at midnight.

With this observation, astronomers could confirm that the calendric system was perfectly in place because there had been no off course after 52 or 104 years of usage. Several authors have proposed dates for the zenith passage of the Pleaides in contemporary times of Sahagún, that is, the sixteenth century (Bernardino de Sahagún 1979: I.327 r.IV, Apéndice),[11] because they assume that day 2 Reed was anchored to the zenith passage of the Pleiades at that time. However, the coupled set of dates was Pleiades based only during the very first Mexica cyclic anniversary, which required "tying" the century of 104 years between years AD 1090 and 1091,[12] so the tradition of celebrating this coupled set of dates was passed on to their Nahua-Mexica descendants until the sixteenth century, even though the Pleiades were not right in the zenith by that time.

The other coupled set, which spans 260 days, is 13 August to 30 April. It is Sun based, since it measures the time elapsed between the Sun's zenith passage on 13 August and its next zenith passage on 30 April. It is also site based, since it can only be followed easily on sites along latitude 15°N through 15°30′N. But all sorts of sacred buildings were erected in order

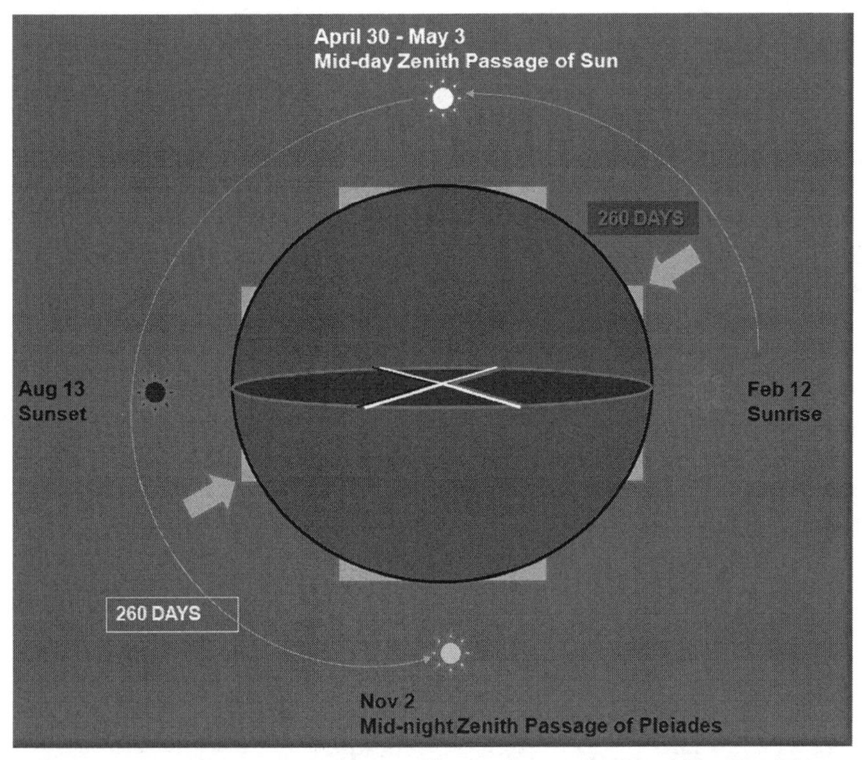

Figure 9.1. Four time–space referents (illustration credit: Geraldine Patrick) The figure depicts four times in a day cycle, four dates within a year, four cardinal segments, and four astronomic referents. The midnight zenith passage of Pleiades applies for a span of about five hundred years starting at the end of the first millennium AD. The 260-day spans are explained in the text.

to follow this period farther north, up to almost latitude 20°N, that is, in Teotihuacan.

Third Concept: Energetic Activation

We have seen up to now that bioculture is ubiquitous; it is linked to a time–space coordinate. Now, if that chronotope is activated with natural energy (mainly in the form of heat or electromagnetic discharge) and/or human energy (mainly as thoughts, words, songs, and music), then biocultural sites acquire a sacred quality[13] and become highly spiritual.

Sacred biocultural sites are considered sites of power and have either a trilogic relation—which means that there is a connection between what is above, what is below, and what is deep within—or a dual relation—which means that as is above, so below. But this relation is not only spatially conceived; it is also mutually conceived in time. Thus, we find a relation of the diurnal world with the nocturnal world but also in the transition phases (i.e., in the transition from night to day and vice versa and even in the transition from the first half of the night to the second half and also between the diurnal halves).

Regularly, a sacred biocultural site on a mountaintop has a corresponding sacred site in the underworld such as a cave, a cavern, a cliff, or a canyon. But it is also connected to a sacred site at the intermediate level, the point of departure to the highlands and to the depths. When this geostrategic location becomes aligned with a certain astronomic time, powerful energies emanate in order to produce a harmonization and a resilient condition for those interacting with them. In short, sacredness of places is infused in biocultural, astronomical, and energetic dimensions, all interlinked.

Long ago, there was a highly developed consciousness about how the spiral movement of the time–space lattice determined sacredness. A specific time–space event was the very first component identified in a natural site that eventually was conceived as sacred. Even today, many places in Mexico are sacred because enchantments occur on 3 May. They may be places rarely visited or known, either in their pristine condition or with some degree of human transformation, but they qualify as sacred because something peculiar happens over and over in a timely pattern.

The astronomical reason underlying the event was once fully acknowledged among expert native people, but it became eroded throughout the colonial and postcolonial periods. This means that the tradition regarding a pilgrimage to a certain place on a certain date and the ceremony held there may have been passed on through generations, but the astronomical referent has been partially lost. It is common to see that such places are now considered sacred due to some Catholic religious asset, making people even more neglectful of the original reason belying the sacredness.

Sacred Biocultural Sites in Mexico

Sacred biocultural sites in Mexico are the dwelling places of biocultures, and they are usually several thousand years old. Many of these sites are unknown to the rest of society due to various reasons. The main one is that adherents of the Christian religion, imposed by Europeans in the sixteenth century, saw, and still see, these places as sites for worshipping the devil. That is the reason why some places, such as some sacred caves, are called "devil's cave." As a result, many native people do not visit or do not know these places.

Another reason is that these sites are hidden from the public. Some custodians keep those sites secret, so outsiders will not destroy them or commit sacrilege there. The importance of these sites lies in the fact that they are vital to cultural cohesion and natural heritage. They are sacred for different purposes; some sites regenerate life, while others are places to give offerings for human and environmental health.

Most of the biocultural sacred sites in Mexico are unknown, and others are being or have already been destroyed. Here we present some of the most known in every category. Various sites are seen more as natural sacred sites and others more as cultural sacred sites. However, we argue that they are biocultural sacred sites because culture and nature are deeply interconnected in practice. Even sacred mountains are biocultural due to the presence of people who venerate them and carry out rituals in such places.

It is necessary to point out that most biocultural sacred sites in Mexico are found in original peoples' territories. Community-based conservation by means of *tequio, nfoxte,* or *faenas*[14] is a normal activity, since according to native philosophy, socioecosystems are compelled to live in a sustainable way.

Thanks to the long resistance of some peoples, in Mexico we find many types of sacred biocultural sites, from mountains to beaches and from caves to deserts, and others. We can find them on mountaintops and volcanoes, in caves, in water springs, in underground caverns, on pilgrimage routes, at sanctuaries, on rocks, in ceremonial centers, in the milpa itself, in some beaches, and even within the perimeter of a complete community or city.

Among the most representative mountains of Mexico, we find the volcanoes in the Central High Plateau. Iztlacihuatl and Popocatepetl, which are perceived as a female-male duality, are found in the Valley of Mexico. Rituals have been carried out in these volcanoes since time immemorial in order to ask for rain on 3 May. Many legends exist around these volcanoes.

Rituals are carried out in the crater of Xinantecatl, a volcano also known as El Nevado de Toluca; offerings are given to the Lagoon of the Sun and the Lagoon of the Moon. These rituals have been happening since pre-Hispanic times; recent archaeological explorations have provided such evidence (Arturo Montero 2010, personal communication).

Today we know, thanks to mountain archaeological research, that cult devotion to volcanoes and mountains is more important than was believed, since many corroborative findings have been registered on almost every mountaintop (Iwaniszewski 1986). Rituals are still done today (Broda 1991) or are being reproduced by those determined to recover traditions. Some sacred mountains are considered places where deities are materialized, so pleadings and offerings are directed to the life nurturer, known as Makihmu among Hñahñu-Otomí and Ometéotl among the Mexica.

There are also mountains where pilgrimage and ceremonies are practiced. In the High Central Mexican Plateau, we find El Cerro de la Campana on the eastern side of the Toluca valley. El Cerrito (Little Mountain) in Mexico State was visited to render cult devotion to Xolotl (vespertine Venus). Pilgrimage is still carried out by walking to its summit, where a Catholic temple is found. Offerings and rituals are still done there; Otomí peoples, Mazahua peoples, and mestizo congregate not merely to attend the religious mass but also to offer dances, songs, and sacred objects around the site. The custodians of the tradition say that everything around the site is sacred. This means that no act of disrespect toward plants, animals, rocks, or soil may occur; if this rule is violated by anyone, he or she may acquire some bad illness.

Also, Chalcantzingo Mountain is a sacred mountain in Morelos that has been offered cult status since Preclassic times. Yet another—of hundreds—is the Peña de Bernal, a place that is visited by people of many cultures who believe it contains a particular energy that revitalizes.

In the current cosmo-vision of native peoples, caves in the mountains allude to the underworld, where sustenance for humanity and ecosystems is produced. But since old times, the underworld has been conceived of as a place of fertility, as an obscure and cold place concerning death. These conceptions are still maintained today (Méndez 2009). So the sacred mountain is a referent to duality, to life and death.

The cave is the sacred place where the duality of evil and good converge. To some, it is the entrance into the underworld, and for others it represents goodness; it is the womb that generates life. Caves represent the entrance into other dimensions where extraordinary happenings occur, and they are also the access to wonderlands, cities, and enchanted places. Rituals are carried out on 3 May in sacred caves, where offerings are made to different entities. These offerings are made to thank Mother Earth for the fruits obtained, to ask for the recovery of the sick, or to venerate Xolotl, evening Venus (the one that can go to the underworld and help the Sun rise victoriously). There are also those who use the sacred sites for witchcraft; however, these are less common.

Caves also enable communication with the deceased (Segura 2005). Sacred caves are found in San Pablito, Pahuatlán, Puebla, Chalma, and Joljá, to name just a few of the many that are found in original peoples' territories. We also find caverns, which are places also used by local people to ask for rain, for health, and for abundance and to give thanks for these as well.

Among Mesoamerican peoples, water springs are sacred places for veneration. Nacelagua water spring is located in the Otomí-Matlatzinca mountain range, a sacred site where Mazahua and Otomí peoples arrive to celebrate life by offering flowers, fruits, bread, food, and resin *copal*. Tolantongo thermal waters have been venerated by Hñahñu-Otomí peoples since time immemorial. Local people have a very organized system based on traditional authorities and *nfoxte* that enables them to administrate several thousand visits per year, while keeping an ecological and hydrological balance as well as the ceremonial spirit of the place.

Cenotes found in southeast Mexico, on the Yucatan Peninsula, are also sacred places containing water. There is a complex system of celebrations around the regeneration of life, and these take place within a calendar that focuses on events regarding agriculture and hunting.

Some lakes are also sacred, like the Zempoala lakes, between Mexico State and the state of Morelos; the Lake of Pátzcuaro in Michoacan is another example of a sacred place where people take offerings to ask for a good harvest, good fishing, or good hunting.

Even the milpa is a sacred site among Mexican peoples, since many cultures in Mexico still carry out rituals in agreement with the agricultural cycle, where the corn seed is consecrated at the beginning of the seeding season and the harvested corn is blessed at the end of the season (Reyes Montes and Albores Zárate 2010).

On the Mexican coastline, there are sacred sites that are still venerated by peoples who inhabit islands or who travel from places far away; for example, the Wixarika peoples from Jalisco and Nayarit go on pilgrimage to the Pacific coast to give offerings to the deity that gave life. The Seri peoples make rituals on Tiburon Island and also on the coast of Sinaloa and Sonora, which are old sacred sites.

There are valleys that are sacred too, such as Tepoztlan. Many people living there believe that there is a special energetic field, so local visits among barrios are carried out and many climb to the mountaintops—particularly to the Tepozteco—to carry out ceremonies.

Sanctuaries are diverse, but they all have connections with sacred sites relevant to ancestors. Most have become syncretic, such as La Villa in Mexico City. Catholic temples are found where old temples used to be located. Most sacred biocultural places are shared by two or more systems of beliefs. Wherever there has been a systematic syncretism since colonial times, the sacred site has suffered from moderate to profound changes in terms of architecture, structure, decoration, and religious motives. Thus, different cultures and beliefs may converge in the same sacred site.

Among the most relevant sanctuaries are Chalma in Mexico State; La Villa, Mexico City; San Juan de los Lagos in Jalisco; Juquila in Oaxaca; and

Santo Niño de Atocha in Zacatecas, to name just a few. Over a million people visit these places every year as an expression of gratitude that they had promised they would offer earlier in the year, or to offer candlelight in memory of the dead, or to obtain renewed energy, or simply as leisure. These places are visited today by pilgrims who travel by foot, on bike, by bus, on horseback, or in personal vehicles.

Among biocultural sacred sites with an enormous biocultural wealth, we find ceremonial centers, which can be either pre-Hispanic or contemporary. Most archaeological sites are being reconceptualized as ceremonial centers, such as Teotihuacan, Mitla, Monte Alban, Palenque, Chichen-Itza, Paquime, Cañada de la Virgen, and Malinalco, among many others. New ceremonial centers have been built in places where ritual ceremonies were carried out according to a ritual calendar. An example is the Otomí Ceremonial Center in Temoaya, which was built in the 1980s. All these sites are currently under the government's administration. In our view, they should be managed by original peoples.

There are places that have the category of *pueblos m*ágicos that in essence have been sacred since very old times; for instance, Malinalco in Mexico State, where the Matlatzinca (of Otomian filiation) and other original peoples used to live. Likewise, there are small communities, such as Pozuelos, from San Juan Chamula, Chiapas, which is considered a mythical place where deities were born; all that surrounds it, like the water springs and the mountains, are considered sacred.

Discussion

We are proposing a new approach to the identification, categorization, and registration of sacred biocultural sites of original and traditional peoples. Our position is that the term "sacred natural site" is subsumed in the term "sacred biocultural site." Two steps occur for a natural site to become a sacred biocultural site. First, as soon as a natural site is conceptualized by a human being, nature becomes cultural. Wherever there are original peoples, a natural landscape becomes a biocultural landscape.

When a site in the biocultural landscape is conceptualized in spiritual symbolic terms and has been energetically activated (by a thunderbolt, or special chants, or even by the solar or lunar position of a particular date), it becomes a sacred biocultural site. This is hard to put into words by local people; it is also difficult for researchers to grasp. So it is problematic to produce protocols that are based on predefined categories for pinpointing SNS. For a start, by definition, if a site has obtained the local sacred nomination, then it is not just natural but biocultural. Scrutiny is needed to identify such places, which is made harder if there is no local participation. Also, categories may vary from bioculture to bioculture. Each site will be a sacred biocultural site as understood by each particular bioculture.

In Mexico, we can assume that a place that is considered "alive" needs to be energetically nourished in times and ways that only some local people know. The point we want to make is that protocols must include local experts and that it is not enough to have those SNS listed in books. Even if they are not, we have to respect that they are in the mindscapes or "heartscapes" of those who are in charge of keeping them imbued with full vitality.

Sacred biocultural sites are places originally signified as valuable in terms of the prevailing time–space philosophy. They eventually became relevant for other reasons that were more easily apprehended by common people, who possess a less complex knowledge permeated by the Catholic religion. In other sites, specific ceremonies or rituals are carried out in coherence with ancestral cosmo-vision and perception of the peoples that are linked to the sites. Such sacred biocultural sites are managed by magic–religious authorities; they motivate peregrinations, prayers, rituals, festivities, cults, and individual or collective sacrifices, providing a sense of identity to the community.

Even though not all places are sacred, all of them have the potential to become so (Petrich 2007). However, there are places that are contaminated or altered to such an extent that it would be almost impossible for them to become sacred. It could happen, though; if they were struck by a thunderbolt, for instance, they could acquire that condition.

Cultural and environmental diversity proposes challenges to environmental sustainability among original peoples in connection with their sacred spaces. A sustainable approach would help a culture and its cosmo-vision to permeate and expand within its territories or zones of concentrated biodiversity where sacred biocultural sites have been preserved.

In the present context, the causes of ecosystem deterioration within original peoples' territories are poverty and histories of political and economic marginalization. Moreover, neoliberal projects of overexploitation of natural systems (ecosystems, geological systems) including illegal bioprospecting with the concomitant subtraction of traditional knowledge and mining, where the principle of prior informed consent is not respected, are all threatening the most important biocultural territories in Mexico, particularly sacred sites. That this is happening obliges policymakers to implement national public policy focused on the protection of sacred biocultural sites with the active participation of experts from local communities of original peoples.

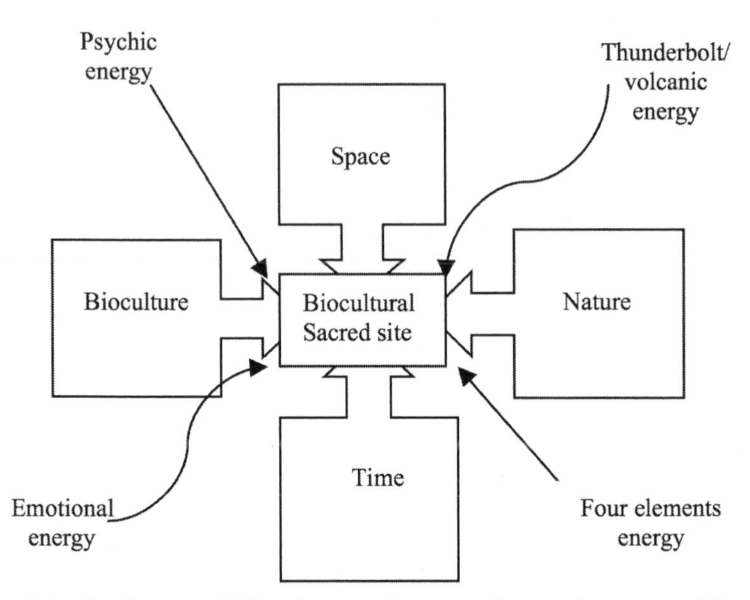

Figure 9.2. Confluence of bioculture and nature, time and space, and human and nature energies

Mindahi Crescencio Bastida Muñoz, Ph.D., is a Scholar in Residence in the Center for Earth Ethics at Union Theological Seminary and a collaborator at the office of UNESCO in Quito for the initiative of Spiritual Reserves of Humanity as well as for the Water and Culture Programme.

Geraldine Patrick Encina, Ph.D., is a Scholar in Residence in the Center for Earth Ethics at Union Theological Seminary. She is an expert on Mesoamerican calendars, astronomy, and conception of time-space.

Notes

1. "Cosmo-vision" refers to the worldview and symbolic representations of a culture. See Barrera-Bassols and Toledo (2005: 26).
2. "Original peoples" is a term used here to refer to a group of people who share an ancestral culture linked to the land and to the sky in symbolic, knowledge-wise, and practical ways. Similar terms are first nations, tribes, indigenous peoples, and native peoples.
3. Note that we are purposely avoiding the use of terms as thing or object, and we use quotation marks for words because the categories belong to the Western optic.

4. Karim-Aly Kassam (2009: 61) says, "Culture emerges from a biological basis—it is an aspect of nature." He discusses Luisa Maffi's (2001) idea that the "species and language binary is an error in logical typing," since "biology and culture are not equal types." That binary, he continues, is a parametric model; this is a "critical error because it reveals a separation between culture and nature in the minds of the proponents."
5. We are seeing this interest in communities in Yucatan (around Valladolid), in José María Morelos, Quintana Roo, and in Toluca Valley, Mexico State.
6. Independent researcher, personal communication, 2012.
7. Thaay Ranchero, Atlixco, Tlaxcala, personal communication, 2012.
8. These portions or segments represent four broad spaces. Two (the eastern and western segments) are drawn by following sunrises and sunsets along the horizon for a whole year. The mirroring spaces between the eastern and western segments become the northern portion and southern portion.
9. Each year of the Mesoamerican calendar is represented by one particular day of the 260-day cycle, which advances together with the cycle of 365 days. In the Otomian and Mexica calendars, this particular day can only be, in four consecutive years: house, rabbit, reed, and flint. A Mexica year is 1 Rabbit because the 364th day of its 365-day cycle is day 1 Rabbit; since the 260-day cycle advances 18,980 positions in order to meet again with the 365-day cycle, we can have a year 1 Rabbit every (18,980/365 = 52) 52 years or its multiples. But any day of the 260-day cycle (for instance, 1 Rabbit) occurs again every 260 days. Hence, if it happens on 10 to 11 February, it will occur again on 28 to 29 October.
10. The slash means transition time between the two Gregorian dates, and it refers to midnight, when day 1 Rabbit commenced.
11. Here follows a free translation of a piece written by Bernardino de Sahagún (1979: I.327 r.IV, Apéndice) in the Códice Florentino: "The biggest count of time that they counted was 104 years, and they called this one century; half of this count, which is 52 years, was called [a] 'bundle' of years. This number of years has been counted since long ago; it is not known when it started, but they knew, by faith, that the world would come to an end at the end of a bundle of years. And they had oracles and prognostics that the movement of the skies would stop, and they took as a signal of the movement the movement of the 'little goats' happening at the night of this celebration, which they called Toxiuhmolopilia; so it happened that the 'little goats' were in the middle of the sky at midnight, with respect to the Mexican horizon. On this night they produced a new fire."
12. The first 104 year bundle ceremony was celebrated between AD 1090 and 1091, according to App. III of the Chimalpahín Codex.
13. "Sacred" is an energetic property, either stable or ephemeral, which is psychologically placed on objects, beings, and places because they are significant for peoples in a collective way and/or because they have obtained natural energetic properties, such as extreme heat or extreme electromagnetism.
14. *Tequio* is the Mexica word for communitarian help; *nfoxte* is the Otomí equivalent; *faena* is the Spanish one.

References

Barabás, Alicia M. 2004. "La Territorialidad Simbólica y los Derechos Territoriales Indígenas: Reflexiones para el Estado Pluriétnico." *Alteridades* 14, no. 27: 105–119.

Barrera-Bassols, Narciso, J A Zinck; E van Ranst. 2003. *Symbolism, Knowledge and Management of Soil and Land Resources in Indigenous Communities: Ethnopedology at Global, Regional and Local Scales.* Enschede: Netherlands: International Institute for Geo-information Science and Earth Observation.

Barrera-Bassols, Narciso and Víctor Manuel Toledo. 2005. "Ethnoecology of the Yucatec Maya. Symbolism, Knowledge and Management of Natural Resources." *Journal of Latin American Geography* 4, no. 1: 9–41.

Berkes, Fikret, Johan Colding, and Carl Folke, eds. 2003. *Navigating Social–Ecological Systems. Building Resilience for Complexity and Change.* Cambridge, UK: Cambridge University Press.

Broda, Johanna. 1991. "Cosmovisión y Observación de la Naturaleza: El Ejemplo del Culto de los Cerros en Mesoamérica." In *Arqueoastronomía y Etnoastronomía en Mesoamérica,* edited by Johanna Broda, Stanislaw Iwaniszewski, and Lucrecia Maupomé, 461–500. Mexico City: Universidad Nacional Autónoma de México, Instituto de Investigaciones.

Bernardino de Sahagún. 1979. *Códice Florentino: Edición facsimilar por el Archivo General de la Nación,* 3 vols., México: Secretaría de Gobernación.

De Sousa Santos, Boaventura, 2010. *Refundación del Estado en América Latina. Perspectivas desde una epistemología del Sur.* Lima: Instituto Internacional de Derecho y Sociedad y Programa Democracia y Transformación Global.

Ingold, Tim. 2000. *The Perception of the Environment: Essays in Livelihood, Dwelling and Skill.* London: Routledge.

Iwaniszewski, Stanislaw. 1986. "La Arqueología de la Alta Montaña en México y su Estado Actual." *Estudios de la Cultura Nahuatl* 18: 249–273.

Kassam, Karim-Aly S. 2009. *Biocultural Diversity and Indigenous Ways of Knowing: Human Ecology in the Arctic.* University of Calgary Press. Available at: http://dspace.ucalgary.ca/bitstream/1880/48645/7/UofCPress_BioculturalDiversity_2009_Contents.pdf (accessed 25 August 2016).

Lenkersdorf, Carlos. 1999. *Los Hombres Verdaderos: Voces y Testimonies Tojolabales.* Mexico City: Siglo XXI.

Maffi, Luisa. 2001. "Introduction: On the Interdependence of Biological and Cultural Diversity." In *On Biocultural Diversity: Linking Languages, Knowledge, and the Environment,* edited by Luisa Maffi, 1–50. Washington, D.C.: Smithsonian Institution Press.

Méndez Pérez, Marcial. 2009. "Reseña del Libro de Daniel Murillo Licea 'Encima del Mar está el Cerro y ahí está el Anjel.' Significación del Agua y Cosmovisión en una Comunidad Tsotsil." *Liminar. Estudios Sociales y Humanísticos* 7, no. 1: 167–169.

Morales Damián, Manuel Alberto. 2014. "Mak, rituales agrarios mayas del fuego y del agua en la Relación de las cosas de Yucatán." In *'Ilu: Revista de Ciencias de las*

Religiones. Available at: http://dx.doi.org/10.5209/rev_ILUR.2014.v19.46616 (accessed 25 August 2016).

Otegui-Acha, Mercedes. 2007. *Developing and Testing a Methodology and Tools for the Inventorying of Sacred Natural Sites of Indigenous and Traditional Peoples*. Gland, Switzerland: Pronatura, Rigoberta Menchu Foundation, IUCN. Available at: http://cmsdata.iucn.org/downloads/otegui_acham_final_report2007.pdf (accessed 25 August 2016).

Otegui-Acha, Mercedes, Gonzales Oviedo, Guillermo Barroso, Martín Gutiérrez, Jaime Santiago, and Bas Verschuuren. 2010. "Developing and Testing a Methodology and Tools for the Inventorying of Sacred Natural Sites of Indigenous and Traditional Peoples in Mexico." In *Sacred Natural Sites: Conserving Nature and Culture*, edited by Bas Verschuuren, Robert Wild, Jeffrey A. McNeely, and Gonzales Oviedo, 209–218. Gland, Switzerland: Pronatura, Rigoberta Menchu Foundation, IUCN.

Patrick, Geraldine. 2013. "Long Count in Function of the Haab and Its Venus–Moon Relation: Application in Chichén Itzá." *Revista Digital Universitaria* 14, no. 5. Available at: http://www.revista.unam.mx/vol.14/num5/art05/index.html (accessed 25 August 2016).

Petrich, Perla. 2007. "Espacios Sagrados entre los Mayas del Lago Atitlán (Guatemala)." *Estudios de Cultura Maya* 29: 141–153.

Reyes Montes, Laura and Beatriz Albores Zárate. 2010. "Agricultura y rituales del tiempo en el Valle de Ixtlahuaca-Jocotitlan, Estado de México." *Colección Documentos de Investigación de El Colegio Mexiquense, A.C.*, no. 134. Available at: http://www.cmq.edu.mx (accessed 25 August 2016).

Segura Romero, Alma Patricia. 2005. "Las Cuevas. Espacio Ritual en el México Colonial. Siglo XVIII." In *Memoria 28, Encuentro de Investigadores del Pensamiento Novohispano*. Universidad Autónoma de San Luis Potosí. Available at: http://www.iifl.unam.mx/pnovohispano/ (accessed 25 August 2016).

Toledo, Víctor M. 1999. *Diversidad Cultural en México*. Reporte para World Wildlife Fund.

Toledo, Víctor M. 2002. "Ethnoecology: A Conceptual Framework for the Study of Indigenous Knowledge of Nature." In *Ethnobiology and Biocultural Diversity: Proceedings of the Seventh International Congress of Ethnobiology*, edited by John R. Stepp, Felice S. Wyndham, and Rebecca K. Zarger. Athens: University of Georgia Press.

Toledo, Víctor M. 2007. *Cultural Diversity, a Key Component of Sustainability*. Background paper requested by UNESCO in preparation for the World Report on Cultural Diversity.

Toledo, Víctor M., editor. 2012. *Red de Etnoecología y Patrimonio Biocultural*. Folleto informativo de CONACYT-México. Mexico City: CONACYT.

Toledo, Víctor Manuel, Pablo Alarcón-Chaires, Patricia Moguel, Abraham Cabrera, Magaly Olivo, Eurídice Leyequien, and Amaya Rodríguez-Aldabe. 2001. "El Atlas Etnoecológico de México y Mesoamerica." *Etnoecológica* 8: 7–41.

Toledo, Víctor Manuel and Narciso Barrera-Bassols. 2008. *La Memoria Biocultural: La Importancia—Ecológica de los Saberes Tradicionales*. Barcelona: Icaria Editorial.

Toledo, Víctor Manuel, Eckart Boege, and Narciso Barrera-Bassols. 2010. "Biocultural Heritage of Mexico: An Overview." *Langscape* 6: 9–16.
Wild, Robert and Christopher McLeod, eds. 2008. *Sacred Natural Sites: Guidelines for Protected Area Managers.* Gland, Switzerland: International Union for Conservation of Nature; Paris: United Nations Educational, Scientific and Cultural Organization.

Chapter 10

New Dimensions in the Territorial Conservation Management in Ecuador

A Brief Political View of Sacred Sites in Ecuador

Xavier Viteri O.

Introduction

Ecuador's constitution protects the country's cultural and natural richness (Sarmiento 1987; Ministerio de Coordinación de Patrimonio 2009), and the Environmental Ministry of Ecuador, created in 1996, promotes the conservation of its rich biodiversity through the creation of fifty protected areas (Ministerio del Ambiente 2014). These are divided into eight different management categories and cover almost 20 percent of the total terrestrial surface of Ecuador's territory. The National Institute of Cultural Heritage (NICH), created in 1978, promotes the conservation of cultural heritage.

The NICH has so far recorded and inventoried more than 168,000 cultural heritage goods (assets of artistic, historic, ethnographic, archaeological, scientific, or technological value, such as artistic works, archaeological sites, archives, and historical documents; see UNESCO 1970 for a similar definition of "cultural property") out of an estimated three million in Ecuador (IDB 2012). The United Nations Educational, Scientific, and Cultural Organization (UNESCO) has also designated several sites in Ecuador as World Heritage Sites, including two as World Natural Heritage Sites (Galapagos Islands in 1979 and Sangay National Park in 1983) and two as World Heritage Sites (the historic center of Quito in 1978 and the historic center of Cuenca in 1999).

One language in Ecuador has also been declared as oral heritage (the Sapara language, in 2008). Also, the Andean Road System, *Qhapaq Ñan*, which spans six countries in South America (Colombia, Ecuador, Peru, Bolivia, Chile, and Argentina), was named a World Heritage Site in 2014. The traditional knowledge of the making of the toquilla straw hat with the fibers of *Carludovica palmata* (mistakenly referred to as a "Panama hat") using the jipijapa technique, was declared an Intangible World Cultural Heritage of Ecuador in 2012. Finally, the marimba music, traditional chants, and dances from the Colombia South Pacific region and Esmeraldas Province of Ecuador were inscribed in 2015 on the Representative List of Intangible Cultural Heritage of Humanity.

In June 2001 the Development Council of Nationalities and Indigenous People of Ecuador developed the System of Indicators of Nationalities and Peoples of Ecuador, which was designed to build indicators that measure wellness or poverty among indigenous people of Ecuador. Núñez, Loaiza, and Cóndor (2003) presented a list of sacred sites within the country that helped to inform this list of indicators. This chapter reviews this publication in the context of developing research on sacred sites in Ecuador and also aims to suggest political possibilities that would allow the incorporation of sacred sites into broader landscapes that could then be managed as either natural or cultural territories. To explore potential ways of doing this, it would be helpful to first summarize what the Ecuadorian constitution says about collective rights and how these are relevant to community management of sacred sites and cultural landscapes in Ecuador.

Cultural and Natural Heritage as a Collective Right

In Ecuador, the concept of "heritage" is tightly connected with the concept of "property." Ownership of natural and cultural heritage assets as "property" in a legal sense is problematic, as they are not just physical entities and locations but also include intangible cultural elements such as knowledge(s), history, and sacredness. In this sense, claims to and "ownership" of cultural and natural heritage occur on several levels simultaneously: individual, collective, national, and global.

Collective rights of indigenous peoples to their heritages include respect for their cultural heritage, access to and control of their sacred and culturally important sites, preservation of their cultural history and memory, conservation and use of the biodiversity within their territories, and protection for the bodies of their ancestors (Núñez, Loaiza, and Cóndor 2003). According to the indigenous cosmological vision, biodiversity is sustained on two principles: (1) world integrity, which results from the cohesion of social and natural worlds and from collective work among humans (called *minga* in the Kichwa [also spelled Quechua, as in Chapter 5] language), and (2) the unity that emerges from diversity, whereby the individual efforts and unique traits of each member of the group (potentially all of creation) contributes to a greater goal that transcends individual aspirations.

This emphasis on social cohesion and shared responsibility is the basis on which they demand their collective rights as a unified political unit. Without this cooperation among individuals with different talents and motivations, the community as a whole will not achieve biocultural cohesion and permanence. That is why American indigenes reject all types of privatization and do not favor trade in wildlife resources, as these are part of a closed socionatural system whose parts cannot be removed without consequence (Weaver 2014).

A high percentage of cultural and natural heritage sites are found within indigenous territories (Lennon and Taylor 2012), and many indigenous areas are now protected as part of a biocultural heritage approach of conservation (Hong, Bogaert, and Min 2014). It is not a coincidence that many areas that are protected by professional environmentalists are actually populated, worked, and conserved by indigenous people, who through their traditions, culture, and history of interactions with the land have for generations passed on respect for the Earth and the idea of humans as custodians of the land (Brown et al. 2005).

This notion is known as *pachamama* in the Andean world and is considered a driving force in the policy-making processes of Andean nations following the new socialism of the twenty-first century (Calisto and Langmore 2014). *Pachamama* roughly translates to "Nature" or "Mother Earth" in English. While there are important and demonstrable links between cultural diversity and biodiversity, these links should not be overstated or oversimplified; nor should the production of new knowledge as a result of syncretism and hybridism be ignored in the face of the devastating loss of much biological and cultural diversity (Brosius and Hitchner 2010). In Ecuador, the inclusion of the rights of Mother Earth (*pachamama*) in the constitution allows for the inclusion of indicators to assess the validity of conservation schemes implemented at the local level.

The System of Indicators of Nationalities and Peoples of Ecuador focuses on the rights of indigenous people, acknowledging the importance of their territories as the vital space where past, present, and future gather together (Núñez, Loaiza, and Cóndor 2003). Social movements in Latin America at large are galvanized around issues of securing land tenure by obtaining legal title, thus territorializing the framework of indigenous identity. As political ecology scholars have shown, the definition of a territory ranges from a place that is actively defended to a space that is guarded against prospective risks, or even to an imaginary space that is not actively contested by others (Escobar 1998).

Thus, indigenous peoples of Ecuador have been actively engaged in demanding property rights and access rights to water, for example, as ways to maintain their cohesive ethnic identity. At present, the constitution of

Ecuador recognizes the need to protect and preserve both culture and nature; it states that Ecuador is a country of "rights and social justice, democratic, sovereign, independent, unitary, intercultural; it is a State of several nationalities and secular."[1] The duties of the state include "to strengthen the national unity in its diversity" and "to protect the natural and cultural heritage of the country."[2]

The rights of people, communities, nationalities, and collectives are recognized and guaranteed in the constitution; these include rights to their knowledge, including their scientific, technological, and ancient and traditional knowledge, and to the genetic resources (biodiversity and agrodiversity) within their territories, thus aligning with the biocultural heritage paradigm (Taylor, Mitchell, and St. Clair 2015). The constitution also protects the rights of indigenous and local communities to "protect, rehabilitate, and promote their ritual and sacred places, including plants, animals, minerals, and ecosystems within their territories,"[3] which also reflects the new thinking about the management of cultural uses in protected areas (Sarmiento et al. 2014).

Ecuador has created and continues to promote the National Plan of Good Living,[4] which is known as *Buen Vivir* in Spanish and is described as "the lifestyle that enables happiness and permanency of cultural and environmental diversity; it is harmony, equality, equity, and solidarity. It is not the quest for opulence or infinite economic growth" (National Secretariat of Planning and Development 2013).

The concepts underlying *Buen Vivir* depart from the traditional development paradigm driven by social actors in Latin America in the last several decades (see Guevara and Frolich, this volume). The National Plan of Good Living, the framework guiding Ecuadorian policy, has twelve objectives. One of them, its fifth objective, refers to building space for social interaction and strengthening national identity, diverse ethnic identities, plurinationality, and interculturality; these latter two goals are aimed at encouraging people to embrace the complexities of identity by acknowledging their identities as both a member of an indigenous group and a citizen of Ecuador.

Ecuador comprises different indigenous nations as well as a mestizo majority, so there is a plurality of nations within Ecuador; the country also harbors a plethora of different cultures that all interact within the national framework of interculturality. This objective proposes strategies to strengthen the plurinational, intercultural identity, by preserving and revitalizing the heritage and diverse collective and individual memories of its people and by promoting cultural entrepreneurship with diverse and inclusive contents (National Secretariat of Planning and Development 2013).

By promoting the respect for nature and cultural values associated with good living, sacred sites can be used as an effective tool to manage the natural and cultural heritage of Ecuador. For example, there are huacas (sacred sites) close to farmscapes where agrodiversity is still important. These sites still sustain cultural and traditional agricultural practices in some Andean indigenous communities, such as those in the Cañar Province, located in southern Ecuador.

Sacred sites are important elements of cultural landscapes, as defined by UNESCO's World Heritage Center (Rössler 2003), created after the Heritage Convention of the United Nations, to which Ecuador is a signatory country. Because of their sanctity and isolation, sacred sites can also help indigenous communities avoid the external pressures of tourism from a globalized world; these pressures threaten traditional lifestyles due to foreign influence and the commoditization of the concept of the noble savage. Sacred sites, through good management of vernacular cultural and natural values, are an important tool in the conservation toolbox of Latin America (Rozzi and Massardo 2011).

Some Views of Sacred Sites in Ecuador

According to Núñez, Loaiza, and Cóndor (2003), one of the goals of the first National Survey of Nationalities and Peoples of Ecuador was to identify culturally important sacred sites. Sacred sites for indigenous people can be found in all elements: water, land, air, and fire; they can embrace all biodiversity and are also found in natural anomalies, minerals found in certain territories, and in heavenly objects. Originally, among Andean indigenous groups, the notion of "the sacred" was expressed with the Castilianized term "huaca," which refers to the idea that sacredness is kept in dark caves or tombs or is hidden in important ceremonial places.

The word "huaca" is derived from the Kichwa word *waka,* which refers to potential communication with a spirit in the underworld; thus, it is often found in tombs, burials, and offerings in deep hollows or built enclaves. In this context, sacred sites can include special mountaintops, nested fields, caves, waterfalls, lakes, forests, and trees whose access is regulated through the years by legends, rituals, and taboos.

For the Chachis, an indigenous group located in northwestern Ecuador, there exist supernatural powers in some mountains and hills that can either cure or cause damage. Shamans can communicate with these souls so that their powers can be used to cause good or harm or lead to certain events or outcomes. Plants and animals are also sacred for some indigenous people, who find links among nature, human beings, and the spiritual world. For the Waorani people, located in Yasuni National Park in eastern Ecuador, plant life is associated not only with growth but also with beauty, power, and energy (Núñez, Loaiza, and Cóndor 2003).

Human beings are associated with leaves; new leaves are associated with newborns, while yellow leaves are associated with old age. Waterfalls have enormous significance for the Shuar people living in central and southeastern Ecuador, because in waterfalls live the souls of their ancestors, the *arutam,* or telluric guardians that protect their homes. The survey by Núñez, Loaiza, and Cóndor (2003) was conducted in 976 communities, and it recorded the existence of 328 places considered sacred or of cultural importance. The highest number (>100 sacred sites) was for waterfalls and watersheds, followed by mountains, hills, cliffs, and caves (<60 sacred sites).

Out of these 328 places registered, 52 percent (mostly waterfalls and watersheds) are under the control of indigenous communities, while others are now controlled by local district governments, at either the county level or the municipal level. The indigenous and local communities with the greatest number of controlled places are the Awa, followed by the Shiwiar or Achwar, and then the Sapara. Other groups, such as the Natabuela, the Saraguro, and the Salasaka, do not have control of any sites, according to Núñez, Loaiza, and Cóndor, 2003.

Cultural Landscapes: A New Concept in Ecuador

No laws protecting specifically ritual and sacred sites, not even places where cultural and natural elements converge, currently exist in Ecuador. There are no categories, under the National System of Protected Areas of Ecuador, that specifically protect sacred sites. Basically all categories protected by the Environmental Ministry of Ecuador are devoted to the protection of natural resources and biodiversity; therefore, more integrated management of territories that includes important values such as culture and heritage is necessary.

To meet this goal of more integrated management that protects both natural and cultural heritage, Ecuador's Ministry of Culture and Heritage proposed the term of "cultural landscape" in its new cultural law that was approved in December 2016. This cultural law considers the articulation of natural and cultural dimensions in the context of a specific territory, strengthening conservation of such cultural assets as biological diversity and agrobiodiversity. For instance, Sarmiento, Cotacachi, and Carter (2008) suggest several sites for the designation of cultural landscapes in Ecuador: (1) the Imbakucha watershed, due to the sacredness of some features, such as the mountains and the big lake; (2) Taita Imbabura and Mama Kutacachi, the two iconic mountains of the basin, with the ceremonial waterfall of Piguchi flowing in between; and (3) the sacred tree, or Pinllucruz of the Otavalo people, on top of the fortress hill, or *pukara,* of Reyloma.

In 2010, the NICH and the Municipality of Urcuquí (located 152 kilometers north of Quito in the Imbabura Province) inventoried all material assets (i.e., archaeological places, colonial and republican buildings, paintings, sculptures, books, and documents) and analyzed the historic and natural context of Urcuquí County. One year later, Urcuquí County was used as a successful model of a territory where the concept of cultural landscape category could be applied. Several local institutions at different political levels (i.e., municipal and parochial governments and the cultural, tourism, and environmental ministries) participated in the construction of the application for the territory to be considered a cultural landscape.

Months later, the now defunct Ministry of Culture and Heritage of Ecuador conducted a study to determine the relevance of the cultural landscape category in Ecuador. The study concluded that it is absolutely relevant to include a new category of management but that there must be a political decision to officially create it.

Ritual and sacred sites would be among the potential categories of Ecuadorian cultural landscape. Article 3 of the present Ecuadorian constitution says that among the principal rights are rights for communities to protect the natural and cultural heritage of Ecuador, allowing for the integration of the Ecuadorian cultural landscape.[6]

Conclusion

The conservation of cultural landscapes requires a redefinition of the current narrative within the new biocultural heritage trope (Brosius and Hitchner 2010). I have argued elsewhere (Sarmiento and Viteri 2015) that biocultural heritage conservation bridges the divide between nature and culture and must take place using examples of Ecuadorian identity and sacred geographies, which until now have been lacking in the official roster of protected areas in the country.

The conditions of preserving both viable landscapes and healthy socio-ecological systems must be met to allow the political processes of indigenous revival to proceed in the context of revaluing indigenous spirituality as a national asset, because it is currently being subsumed by dominant Western religious views on sacred sites. The declaration of Ecuadorian cultural landscape will require political action, especially in areas where sacred sites and ritual practices are located and performed. Within the matrix of sacredness, sites that are identified as the most sensitive should be prioritized for immediate protection. Successful implementation of this conservation category can become an effective management category that supports and aligns with Ecuador's framework of *Buen Vivir.*

Xavier Viteri O. M.Sc., is an independent consultant on conservation biology, an expert on biological corridors, and a specialist on natural heritage at Ecuador's Ministry of Culture and Heritage. He is also a guest professor at Simón Bolívar Andean University for the Master's Program in Climate Change and Environmental Business.

Notes

1. Politic Constitution of Ecuador's Republic 2008, Title 1, Article 1.
2. Politic Constitution of Ecuador's Republic 2008, Title 1, Article 3.
3. Politic Constitution of Ecuador's Republic 2008, Chapter 4 ("On the Communities and Nationalities"), Article 57, Number 12.
4. National Secretariat of Planning. 2013. *National Development Plan / National Plan for Good Living 2013-2017.* Quito: National Secretariat of Planning and Development.
5. Politic Constitution of Ecuador's Republic 2008, Title 2, Article 21.
6. Politic Constitution of Ecuador's Republic 2008, Chapter 4, Article 57, Number 13.
7. Politic Constitution of Ecuador's Republic 2008, Title 5, Chapter 4, Article 264.

References

Brosius, J. Peter and Sarah Hitchner. 2010. "Cultural Diversity and Conservation." *International Social Science Journal* 61, no. 119: 141–168.

Brown, Jessica, Nora Mitchell, and Michael Beresford, eds. 2005. *The Protected Landscape Approach: Linking Nature, Culture and Community.* Gland and Cambridge: IUCN—World Conservation Union.

Calisto, Martín and John Langmore. 2014."The Buen Vivir: A Policy to Survive the Anthropocene?" *Global Policy* 6, no. 1: 64–71.

Carter, Lee Ellen and Fausto O. Sarmiento. 2011. "Cotacacheños and Otavaleños: Local Perceptions of Sacred Sites for Farmscape Conservation in Highland Ecuador." *Journal of Human Ecology* 35, no. 1: 61–70.

Escobar, Arturo. 1998. "Whose Knowledge, Whose Nature? Biodiversity, Conservation and the Political Ecology of Social Movements." *Journal of Political Ecology* 5: 53–82.Hong, Sun-Kee, Jan Bogaert, and Qingwen Min, eds. 2014. *Biocultural Landscapes: Diversity, Functions and Values.* New York: Springer.

Hong, Sun-Kee, Jan Bogaert, and Qingwen Min. 2014. *Biocultural Lands: Diversity, Functions and Values.* Springer Netherlands.

IDB [Inter-American Development Bank]. 2012. "Ecuador Will Strengthen Cultural Heritage Protection with IDB Support." 11 January 2012. Available at: http://www.iadb.org/en/news/news-releases/2012-01-11/ecuador-will-strengthen-cultural-heritage-protection,9806.html (accessed 20 July 2014).

Lennon, Jane and Ken Taylor, eds. 2012. *Managing Cultural Landscapes.* Oxon: Routledge.

Ministerio de Coordinación de Patrimonio. 2009. *Agenda del Consejo Sectorial de Política de Patrimonio 2009–2010.* Quito: Gobierno Nacional del Ecuador and Ministerio de Coordinación de Patrimonio.

Ministerio del Ambiente. 2014. "¿Qué son las Áreas Protegidas?" Available at: http://www.ambiente.gob.ec/areas-protegidas-3/ (accessed 21 August 2014).

National Secretariat of Planning. 2013. *National Development Plan / National Plan for Good Living 2013–2017.* Quito: National Secretariat of Planning and Development.

Núñez, Martha, Alicia Loaiza, and Jorge Cóndor. 2003. *Sacred Sites in the Territories of Nationalities and Peoples of Ecuador: An Advance for its Focalization.* SIISE—Sistema Integrado de Indicadores Sociales del Ecuador. Quito: Secretaría Técnica del Frente Social.

Rössler, Mechtild. 2003. "Linking Nature and Culture: World Heritage Cultural Landscapes." In *Cultural Landscapes: The Challenges of Conservation,* edited by Mechtild Rössler, 10–15. World Heritage Papers No. 7. Paris: UNESCO World Heritage Centre.

Rozzi, Ricardo and Francisca Massardo. 2011. "The Road to Biocultural Ethics." *Frontiers in Ecology and the Environment* 9: 246–247.

Sarmiento, Fausto O. 1987. *Antología Ecológica del Ecuador. Desde la Selva hasta el Mar.* Casa de la Cultura Ecuatoriana. Quito: Museo Ecuatoriano de Ciencias Naturales.

Sarmiento, Fausto O., Edwing Bernbaum, Jessica Brown, Jane Lennon, and Sue Feary. 2014. "Managing Cultural Features and Uses." In *Protected Area Governance and Management,* edited by Graeme L. Worboys, Michael Lockwood, Ashish Kothari, Sue Feary, and Ian Pulsford, 685–714. IUCN E-Book. Canberra: Australia National University Press.

Sarmiento, Fausto, César Cotacachi, and Lee Ellen Carter. 2008. "Sacred Imbakucha: Intangibles in the Conservation of Cultural Landscapes in Ecuador." In *Protected Landscapes and Cultural and Spiritual Values,* edited by Josep-Maria Mallarach, 114–131. Values and Protected Landscapes and Seascapes, vol. 2. Heidelberg: IUCNGTZ and Obra Social de Caixa Catalunya, Kasparek Verlag.

Sarmiento, Fausto and Xavier Viteri O. 2015. "Discursive Heritage: Sustaining Andean Cultural Landscapes amidst Environmental Change." In *Cultural Landscapes: Challenges and New Directions,* edited by Ken Taylor, Nora Mitchell, and Archer St. Clair, 309–324. New York: Routledge.

Taylor, Ken, Nora Mitchell, and Archer St. Clair, eds. 2015. *Cultural Landscapes: Challenges and New Directions.* New York: Routledge.

UNESCO [United Nations Educational, Scientific and Cultural Organization]. 1970. "Convention on the Means of Prohibiting and Preventing the Illicit Import, Export and Transfer of Ownership of Cultural Property 1970." Available at: http://portal.unesco.org/en/ev.php-URL_ID=13039&URL_DO=DO_TOPIC&URL_SECTION=201.html (accessed 21 August 2014).

Weaver, Jace. 2014. *The Red Atlantic: American Indigenes and the Making of the Modern World, 1000–1927.* Chapel Hill: University of North Carolina Press.

Chapter 11

Sustainability and Ethnobotanical Knowledge in the Peruvian Amazon

New Directions for Sacred Site Conservation and Indigenous Revival

Fernando Roca Alcázar

Introduction

The ethnic groups of the Amazon Jungle have, for many generations, made abundant use of tropical forest flora for medicine and food as well as for ritual and divination purposes. In this chapter, I will use the example of two upper Amazonian cultures, the Awajún (formerly called Aguaruna) and the Wampis (formerly called Huambisa or Shuar Wampis), to demonstrate the applications of ethnobotanical knowledge to sacred site conservation and indigenous revival within the larger frame of sustainability in Peru. The term "sustainability" can have many meanings. Here I choose one that differentiates between ecological sustainability, marked by resilience to disturbances, and its use in a socioeconomic context, which is linked to long-term access to ecosystem services:

> From an ecological viewpoint, sustainability refers to the persistence of a self-sustaining ecosystem that has sufficient resilience to recover to an intact state, should it suffer from disturbance. In a socioeconomic context, sustainability is the application of sound ecological principles in order to derive ecosystem services on a continuing basis without causing harm to ecosystems that provide these services. (Clevel and Aronson 2013: 13)

Sustainable development is premised on the recognition that human activities affect elements of the natural environment, including air, water, forests, plants, animals, and other natural resources within the context of contentious power relations of the globalized market (Zimmerer 2006).

Use of ecosystem services can take many forms, including large development projects such as infrastructure construction, the use of petroleum and gas resources, mining and other extractive activities, the timber industry, industrial fisheries, industrial agriculture, and large-scale cattle ranching. However, it can also take the form of local or community-scale projects that affect limited areas with their corresponding natural resources (Brannstrom and Vadjunec 2014).

In any of these cases, traditional ecological knowledge (TEK) and new technologies can both contribute to long-term economic viability. Today, there is wide recognition of the need to mitigate anthropogenic impacts on the biosphere. While nature does not need human beings to continue its evolutionary processes, human beings do need nature to continue living on this planet. At the same time, we are the only beings on Earth that can intentionally influence nature. Studies of natural history have shown that, over long time frames, nature can absorb the impacts from many, though not all, human activities (Balée 1994). Balée's contributions to showing the influence of ancient peoples on tropical forests of the Amazon are the basis for a new understanding of tropical rain forests from the prism of historical ecology (Balée 2014), and this theoretical framework has contributed greatly to the redefinition of "pristine" jungles as cultural landscapes (Rostain 2012; Denevan 2014; McMichael et al. 2014).

Over long periods of time, natural resources and associated ecosystem services have, in many cases, been able to recover from exploitative human activities such as indiscriminate hunting and fishing, deforestation, and pollution of soil, water, and air. Such recoveries have usually occurred in areas of lower human population densities. We are currently approaching a breaking point in this relationship between humans and nature, even in such isolated areas as the Amazonian hinterlands. As the world's population increases, so too does the pressure that humans exert on the environment.

To improve the odds of long-term sustainability (both ecological and socioeconomic), we must use appropriate technologies and collective wisdom to manage and reduce this anthropogenic impact on tropical rainforest ecosystems. Otherwise, we will jeopardize future generations, limiting their quality of life and even endangering the possibility of living on this planet. Because of the importance of traditional ethnobotanical knowledge to maintaining indigenous people's sustainability in the rainforests, the relevance of conserving sacred natural sites (SNS), which harbor great biodiversity, has become an important focal point for conservation science and ethnographic research efforts.

Case Study: The Amazonian Rain Forest

As humans use natural ecosystem services, these become transformed into social environmental services. These social environmental services can take many forms, and their impacts on nature can be very diverse (MEA 2005). This chapter examines one of these possible forms in a specific ecosystem and its local actors: the Amazonian tropical rainforest and some of its inhabitants, the *bosquesinos,* literally the "people of the forests," who are known to outsiders as the original inhabitants or indigenous Amazonian groups (Gasché and Vela-Mendoza 2011: 31).

The original Amazonian people, or *pueblos originarios amazónicos,* have developed highly specialized interactions with their environment over millennia, and these interactions have resulted in a complex and nuanced body of TEK. This knowledge has led to creative and innovative uses for natural resources and simultaneous preservation of the ecosystem and a high quality of human life. Sacred sites are an integral part of this knowledge. The *bosquesinos* have created well-developed taxonomies of flora and fauna and amassed detailed knowledge about the characteristics and habits of the living beings within those ecosystems.

This knowledge is evident in their oral narratives, which are transmitted from generation to generation. Their mythic traditions also speak about sacred places and about other natural elements, including soil qualities, the landscape and its morphology, ecological components of different forests, and plants that are edible, medicinal, or hallucinogenic. In addition, their myths also tell stories about how to adapt their livelihood strategies to changes in weather and how to survive in extreme conditions or in times of war against other surrounding indigenous groups.

In many Amazonian indigenous groups, biodiversity conservation strategies and sustainable use of natural resources are directly correlated to an appropriate number of people that can be supported by the natural environment (Otsuki 2013). This ideal demographic density, combined with traditional knowledge and sustainable use of natural resources, was compatible with livelihood strategies that resulted in low impacts on the natural environment and therefore maintaining SNS as core protected places within their territories.

This chapter focuses on the Awajún and Wampis (a subfamily of the Shuar Jívaro group) groups, both of which are members of the larger Jívaro ethnolinguistic group. These two large, closely related indigenous groups live between Peru and Ecuador at the eastern slope of the Andes, in the interface between the Amazonian rain forest and the Andes mountains. Their settlement patterns are characterized by small familial groups, generally living near rivers, streams, or rapids in isolated houses, which are usually strategic places from which it is possible to control a territory or access to a river or ravine.

Warfare has traditionally been a principal cultural characteristic of the Jívaro group, and the strategies for waging war were very important coping mechanisms (Harner 1984); all the Jívaro subgroups were enemies, and war was very common between them (Sarmiento 1961; Guallart 1990; Guallart 1995). Their houses showcase the constant reality of warfare, as they have two entrances, allowing the possibility of escape in case of an attack. The following section provides a short ethnographic description of the Awajún and Wampis people, giving context for later discussion of the cultural significance of their SNS.

Principal Awajún and Wampis Cultural Characteristics

On the south side of the main Marañón River, there are two secondary rivers, Chiriaco and Nieva. On the Marañón's north bank there are two secondary rivers, Cenepa and Santiago, which are both located in the headwaters of the Marañón watershed of highland Amazonia, which includes territory of both Ecuador and Peru. The Santiago, whose headwaters are in Ecuador, is the biggest river after the Marañón in this area, and the Huambisa people are settled just on the Peruvian side of the river (Guallart 1995; De Saulieu 2006).

The most accurate data regarding the origin of these people suggest that they have been living in this region for about one thousand years (Guallart 1990). The traditional Aguaruna house has an elliptic design, has two entrances, and is surrounded by domestic plantations or cleared areas with no encroaching tropical rainforest. These houses used to be organized internally in two parts: one linked to the masculine cosmology and the other to the feminine, a configuration that has been interpreted as a symbolic differentiation between formal and informal activities (Brown 1985).

In the past, the head of each familial domestic unit was a young man with his wife or wives and children; today, however, these characteristics are changing very quickly. Polygamy is still the dominant family structure in some areas, but monogamy is rapidly becoming more commonplace. The head of the family's daughters' husbands could be included in this unit, at least during the first years of marriage, according to the will of the father-in-law. The houses, linked by kinship relations, were disseminated over a common territory. The total population of these nuclear families could range from 80 to 120 inhabitants, although some lived far away from the others in a settlement pattern called an "endogamic nexus" (Descola 1982: 305). Brown (1985: 127–128) and Descola (1982: 305) show that in the gender-related work–labor division, men are essentially "predators of nature" and women are "the cultural transformers of nature." It is important to understand this distinction, as it is relevant to the different types of relations that men and women develop with nature.

Water is sacred for many indigenous groups all over the world (Altman 2002). Although many Amazonian ethnic groups have origin myths in which people emerge from water, this is not the case with the Awajún Jívaro group. The Awajún oral tradition says that there was a mythic war between the beings of the water and the beings of the land. The Awajún came from the land, not from the water, and in the last big battle against the spirits and beings of the water, they were victorious (Chumap-Lucía and Garía-Rendueles 1979). Despite active engagement with the economic activity of the dominant society, through radiophonic schools and cattle ranching, these groups are still able to maintain traditional sacred places (Rudel, Bates, and Machinguiashi 2002).

The principal sacred sites of the Awajún and Wampis are the waterfalls, which are contact places between humans and the spirits of the forest (Harner 1984; Bennet 1992). Also sacred are their cemeteries, where strict rules dictate treatment of the bodies of the dead (Descola 1982). However, cemeteries have changed, as modern Christian rites are replacing traditional mortuary rituals. Few people now prepare the dead bodies in the traditional way, which included wrapping them in bark to be hung in tall trees. At present, it is not uncommon to see Awajún and Wampis people digging graves in which to inter the deceased. However, waterfalls and cemeteries have maintained their sacred nature in spite of wider engagement with other cultures and worldviews.

In addition, as in many other groups throughout the pan-Amazon territory, sacred sites and special pilgrimage places are divined by shamans via *Ayahuasca* (*Banisteropsis caapi*) hallucinogenic ceremonial rites (Luna 2003). The Awajún and Wampis people also engage in other practices of ritual warfare and agricultural production. These practices are summarized in jungle myths (Sarmiento 1970; Chumap-Lucía and García-Rendueles 1979), which have been collected as part of the extensive study of the custom of creating shrunken heads in a ritual ceremony named *tzantza* (or *tsantsa* in the Jívaro language), as well as other Jívaro cultural practices (Sarmiento 1961; Chumap-Lucía and García-Rendueles 1979). More recently, the Missouri Botanical Garden published the two-volume *Guide to the Cenepa River Flora,* in both English and Aguaruna, the toponymy section of which lists new names for some geographical places considered to be sacred (Vásquez, Gonzáles, and Van der Werff 2010).

With the passage of the Peruvian Law of Native Communities and Agrarian Development of 9 May 1978, the settlement patterns were altered, forcing many of the Awajún and Wampis to concentrate in one single place, giving rise to the *comunidad nativa,* or native community. This arrangement was necessary so that indigenous people could be given titles to their lands within this new judicial and cultural settlement pattern of communal property.

The consequence was a gradual change in the biodiversity conservation strategies, the sustainable use of natural resources, and the appropriation of the territories with these new judicial as well as cultural boundaries (Bennett 1992). Pressure on the territory became more acute, and new lands were opened up for chacra, or household cultivation plots. In the same way, new commercial plantations appeared, mainly banana plantations. An increase in hunting and fishing led to the disappearance of many kinds of mammals, birds, and fish. Gradually, forest resources dwindled around both domestic and commercial plantations near those communities.

Areas near the Cenepa River quickly revealed this degradation around inhabited places (middle and lower stream), while the Santiago River showed higher biodiversity because this area experienced less farmscape transformation and reduced settlement pressure. The same holds for the flora, as original peoples gathered wild fruits to eat and other plants to use as raw building materials or for tools for work and defense (Guallart 1995).

Before the establishment of Peruvian state institutions, the Awajún and Wampis people did not live in large settlements. Today, more than two thousand inhabitants live in native Awajún communities, such as that of the Kuzú-Kubeim, who live near the confluence of the Comaina and Numpatkem rivers in the Cenepa River basin and that of the Wampis native community of Yutupis in the Santiago watershed. Recent studies of the languages and demography of Peruvian Amazon indigenous peoples affirm that some groups, including the Awajún, Wampis, and Ashaninka, are increasing in number (Chirinos Rivera 1998). In the future, these groups are likely to merge, with their demographic characterization consolidated. For the Jívaro Achuar on the Ecuadorian side, changes linked to the traditional settlement pattern began in the 1970s with the introduction of cattle and the construction of airstrips as a result of missionary action (Descola 1982: 314–316).

These changes among the Awajún and Wampis inhabitants began in about 1965 with the creation of, principally Catholic, primary schools and continued with the promulgation of the previously mentioned Law of Native Communities and Agrarian Development in 1978. In order to have primary schools, the Awajún and Wampis families, who were traditionally enemies, had to be grouped together in small villages in order to meet the requirements for minimum numbers of students. Later, to give them property titles to the land, the Peruvian government required a certain number of inhabitants living in a village. The bigger the village, the greater the area of land that would be communally titled. This law forced the two ethnic groups to live in these new concentrations, not in the traditional Jívaro manner.

These two stages marked the beginning of changes in kinship relationships, settlement patterns, strategies for biodiversity conservation, and territorial use and home range demarcation. Traditional knowledge began to undergo intensive changes and adaptations as a result of this new situation. Taboo mammals like red deer, called *japa* in Jívaro (*Mazama americana*), began to be hunted by young Jívaro people. The Jívaro people consider jaguars and deer taboo because after someone passes away, his or her spirit begins a journey, and the spirit inhabits these animals during the first stage of this journey. Deer are no longer taboo for young people, but jaguars and all other felines continue to be considered taboo animals for both young and old people.

In a traditional society like those of the Awajún and Wampis, who were isolated from Peruvian society until ninety years ago, concepts like market economics, land accumulation, and their respective economic values, did not exist. But in the last two decades of the twentieth century Western acculturation has become acute, with the exponential demographic growth of the indigenous people and the stronger presence of the Peruvian military, police, and civil society.

In light of this situation, it is imperative to conduct research on traditional knowledge and oral tradition in order to archive past practices as well as current and ongoing cultural changes. Research demonstrates that the Awajún and Wampis still recognize their sacred sites in the waterfalls and cemeteries of the Upper Marañón and that there is an indigenous revival in the face of impending globalization.

Relationships between Nature and Ceremonies

In order to adequately respond to new knowledge about rapid global changes, local knowledge and oral tradition must introduce new parameters adapted to changing demographic conditions and develop new strategies that are culturally different from those practiced many decades ago. It is imperative to prevent ancestral traditional knowledge from disappearing; however, TEK also has to articulate new and emergent social and cultural elements.

Several examples of modern improvements to traditional ways that could benefit people and keep ancient cultures alive include the domestication of wild animals for protein provisioning, agronomic improvements that increase crop yield for family use or commerce, the establishment of water channels (potable water, if possible), and improved hygiene, which could include incorporating the construction of latrines into the design of the new villages. Also, traditional and natural construction materials cannot be found as easily as in the past, when they could simply be found in the rainforest. These resources are dwindling, and continued use will produce a rapid exhaustion of these resources from the rainforest.

This situation becomes more complex when confronting other external factors such as the arrival of low-wage laborers coming from the northern Andean sierra or the northern Peruvian coast (Salisbury and Weinstein 2014). These laborers create new villages, and the influx of outsiders to this region also leads to illegal mining, deforestation due to illicit trafficking of forest resources, and the establishment of poppy or coca plantations for the production of heroin and cocaine for both personal consumption and narcotics trafficking. In addition to this unorganized and informal colonization by mestizos and others who are not *bosquesinos,* legal and formal mining concessions, as well as oil and gas projects sanctioned by the Peruvian government, can also have drastic and long-lasting impacts on the local environment.

These latter projects may develop environmentally friendly strategies if they are guided by a sense of social responsibility; they may attenuate the impact that these industries have on the environment where they work. Nevertheless, in these cases, we must take into consideration the International Labour Organization's 169 decree (known as ILO Convention No. 169), a legally binding international instrument, ratified so far by twenty countries, that ensures rights for indigenous and tribal peoples to self-identification, protection from discrimination, and free and open participation in decisions that affect their livelihoods and well-being, among other things (ILO 1989). Although detailed discussion of this decree lies beyond the scope of this brief exposition, it is important to note here that it introduces new opportunities for indigenous communities to preserve their past and guide their future.

The dilemma of sustainable development then arises: is it possible to establish a link between scientific knowledge and the reformulation of traditional knowledge in this new dynamic context? Is it possible to speak about sustainable development that takes into account exogenous ideas and is based on endogenous traditions?

In Search of Sustainable Development

Several experiences in the Peruvian Amazon may shed light on the possibility of combining tradition with modern science. It is interesting to note that much of this syncretism arose in an endogenous manner, highlighting both the necessity for and the ability of indigenous peoples to tackle modernity in their own way.

These indigenous groups have, for example, designed innovative and adaptive strategies for the use of palm trees, which are a very important cultural resource. There are indigenous communities that have begun to specialize in growing palm tree leaves, which are very important for thatching their houses, in particular, the *yarina* (*Phytelephas macrocarpa*), the tagua, or "vegetal ivory" palm, in the Upper Marañón, Condorcanqui Province, Amazonas region. These communities have planted large areas of *yarina,* and the production and harvesting of leaves provide them income. They sell the leaves to other communities or trade them in exchange markets. Some villages have large populations of *aguaje* palm trees (*Mauritia flexuosa L.f.*), which they manage sustainably. The fruit of this palm is very nourishing and, as a consequence, it is in high demand. Others use their *chambira* (*Astrocarium chambira* Burret) plantations to produce leaves; the sword leaves are commonly used to weave many products of cultural use (including, but not limited to, baskets, small bags, hammocks, and fishing nets), and the tasty fruits of these trees are also a source of income.

We can also find mixed development formulas in which local people have designed strategies with the input of private institutions, nongovernmental organizations, or government institutions. One example of this is the sustainable use of the *irapay* or *palmiche* (*Lepidocaryum tenue* Martius) in the Loreto region (Mendoza 2007), where local knowledge and techniques of sustainable development are compatible with scientific knowledge. The local inhabitants export very attractive baskets and other products made with the leaves of this pleasing looking palm. Other examples include projects that involve fish farming, the domestication of wild Amazonian mammals, tropical fruit production, the organic production of the heart of palm trees (*Bactris gasipaes*) or cacao (*Theobroma* spp.), and small forestry development. All these projects are sponsored by the IIAP (Instituto de Investigaciones de la Amazonía Peruana, or Research Institute for the Peruvian Amazon).

An interesting example of benefits, but not without some contradictions, is handicraft production of commercial pottery pieces made by the Shipibo indigenous people (Mundo Shipibo 2006) that are now popular in exclusive shops and expensive art studios in major capital cities. The Shipibo receive income from selling their handicrafts, but the primary motivation for producing these handicrafts is not to get money. Rather, the production of handicrafts puts the craftsman or craftswoman in contact with the cosmos. The commodification of these handicrafts can lead to the loss of cultural meaning of this process. These types of projects are run alongside infrastructure programs for potable water, latrines, and health campaigns as mentioned, as these are staple prescriptions of any development program in the area. However, only those programs that have incorporated local cultural elements in different ways have really been effective.

Palliatives and Long-Term Proposals

The traditional relationship that the *bosquesinos* established with the rainforest was based on sharing, because they felt that they were part of the rainforest and not the absolute owners of it (Goodman and Hall 1990). This relationship was translated into "sustainable" ways to use the ecosystem services for their benefit, while protecting their environment: rules for hunting and fishing with bans during some periods of the year when terrestrial and aquatic fauna are breeding, planned distribution of the territory to keep some areas cleared of encroaching tropical rainforest, the establishment of plantations for domestic use, and standardized methods for the small-scale commercial production of bananas and manioc.

In the case of many Peruvian Amazonian indigenous groups, the demographic increase is obvious, as evident in the latest census numbers. Conversely, other groups are diminishing at an alarming rate. In the case of the groups with demographic increases, changes in the patterns of land use and land cover are guaranteed. Many of the proposals mentioned above can work as temporary answers, or palliatives, in exchange for deeper changes that can only be confirmed with longer time scales. The risk of the loss of traditional knowledge in the young indigenous generations is real. Those young people who have received a formal education, who are perhaps bilingual but not bicultural, are those most threatened. If they do not have a basic understanding of traditional knowledge, how can they invigorate native culture or revive ancient practices to make changes that once linked humans and nature, while grappling with new concepts learned in school?

In the Peruvian Amazon region, *bosquesinos* need to be optimistic but not naïve in building capacity that enables people to continue their traditional lifestyles in the ways that they choose. To elaborate on past practices and to process new ways of coping with change are inherent characteristics of *bosquesinos*. For instance, in the 1980s I witnessed indigenous people reject the *cuy* or guinea pig (*Cavia porcellus*) as a new foodstuff in the Upper Marañón area because it looked like a rat and originated in the Andes mountains. However, today we can see guinea pigs in many Awajún and Wampis homes as part of the local diet and also as pets.

In the same way, fish farming in small lakes or lagoons and the domestication of wild Amazonian mammals, such as agouties (*Dasyprocta* spp.), capybaras (*Hydrochoerus hydrochaeris*), peccaries (*Tayassu pecari*), and even tapirs (*Tapirus terrestris*), show us how these changes in cultural processes can be implemented successfully. They are, in fact, strategic cultural adjustments to cope with the pressures of a Westernized economy in the tropical rainforest. These changes are made to maintain a balanced diet and, at the same time, to permit a repopulation of the biota of the Amazonian cultural landscapes.

Conclusion

Alternative strategies and techniques for land use and management under the pressure of the new variable of increased population, which is exponential among some indigenous groups, can permit native populations to continue to increase their numbers without damaging the Amazonian rainforest ecosystem (Nikolakis and Innes 2014). Furthermore, these new alternatives should be incorporated gradually into their TEK of animals and plants, further linking ethnobotanical knowledge with the tropical environment, as demonstrated with the new commercial uses of different native palm trees.

Oral tradition, therefore, will add to this process of cultural affirmation and indigenous revival, because it merges traditional knowledge with new proposals and knowledge that is continually being renewed as the Awajún and Wampis face globalization. Nevertheless, sacred sites like waterfalls and ancestral cemeteries retain their cultural significance, and people today still respect them as they did before. Indigenous revival is an ongoing process, as local communities continue to elaborate this synthesis between the modern and the traditional, the new and the old.

There still remains the question about safekeeping the ancestral knowledge of the rainforest and its resources: will TEK continue to be transmitted from generation to generation, allowing for the sustainable management of nature by humans as they interact with the ecosystem? And, more importantly, will the advancement of SNS conservation allow for a dialogue between modernity and ancestrality that will encourage new kinds of knowledge to syncretize with the old ones? While there is uncertainty, I believe that there is the potential for this syncretization and that this dialogue will lead not just to palliatives but to long-term cultural restructuring in which native groups continue to utilize the flexibility that has allowed them to adapt to change throughout the past and across many generations.

Fernando Roca Alcázar, S.J., Ph.D., is a principal professor at Pontifical Catholic University of Peru (PUCP), Lima, Peru. He is a social anthropologist and ethnobotanist as well as the director of the PUCP Ethnobotanical Garden and an expert on the botanical use of Arecaceae by indigenous people.

References

Altman, Nathaniel. 2002. *Sacred Water: The Spiritual Source of Life.* Mahwah, NJ: Paulist Press.

Balée, William. 1994. *Footprints of the Forest: Ka'apor Ethnobotany—The Historical Ecology of Plant Utilization by an Amazonian People.* New York: Columbia University Press.

———. 2014. "Historical Ecology and the Explanation of Diversity: Amazonian Case Studies." In *Applied Ecology and Human Dimensions in Biological Conservation,* edited by Luciano M. Verdade, Maria Carolina Lyra-Jorge, and Carlos I. Piña, 19–33. New York: Springer.

Bennett, Bradley C. 1992. "Plants and People of the Amazonian Rainforests." *BioScience* 42, no. 8: 599–607.

Brannstrom, Christian and Jacqueline M. Vadjunec, eds. 2014. *Land Change Science, Political Ecology and Sustainability: Synergies and Divergences.* New York: Routledge.

Brown, Michael. 1985. *Tsewa's Gift. Magic and Meaning in an Amazonian Society.* Smithsonian Series in Ethnographic Enquiry. Washington, D.C. and London: Smithsonian Institution Press.

———. 2014. *Upriver: The Turbulent Life and Times of an Amazonian People.* Cambridge, MA: Harvard University Press.

Chirinos Rivera, Andrés. 1998. "Las Lenguas Indígenas Peruanas Más Allá del Año 2000: Una Panorámica Histórica." Revista Andina 32, no. 2: 453–79. Available at: http://red.pucp.edu.pe/ridei/files/2011/08/203.pdf (accessed 25 August 2016).

Chumap-Lucía, Aurelio and Manuel García-Rendueles. 1979. *"Duik Muun ..." : Universo Mítico de los Aguaruna.* Serie Antropológica, vol. I. Lima, Peru: Centro Amazónico de Antropología y Aplicación Práctica.

Clevel, Andre, James Aronson, and Society for Ecological Restoration. 2013. *Ecological Restoration. Principles, Values and Structure of an Emerging Profession,* 2nd ed. London: Island Press.

De Saulieu, Geoffroy. 2006. *Una Introducción a la Amazonía Ecuatoriana Prehispánica.* Quito: Editorial Abya-Yala.

Denevan, William M. 2014. "Estimating Amazonian Indian Numbers in 1492." *Journal of Latin American Geography* 13, no. 2: 207–221.

Descola, Philippe. 1982. "Territorial Adjustments among the Achuar of Ecuador, Human Societies and Ecosystems." *Social Science Information* 212: 301–320.

———. 1986. *La Nature Domestique: Symbolisme et Praxis dans l'Écologie des Achuar.* Paris: Editions de la Maison des Sciences de l'Homme.

Gasché, Jorge and Napoleón Vela-Mendoza. 2011. *Sociedad Bosquesina,* vol. I. Iquitos, Peru: Instituto de Investigaciones de la Amazonía Peruana; and Kyoto, Japan: CIES.

Goodman, David and Anthony Hall. 1990. *The Future of Amazonia: Destruction or Sustainable Development?* New York: Macmillan.

Guallart, José María. 1990. *Entre Pongo y Cordillera.* Lima: Centro Amazónico de Antropología y Aplicación Práctica.

———. 1995. *La Tierra de los Cinco Ríos.* Lima: Pontifical Catholic University of Peru, Instituto Riva Agüero, Banco Central de Reserva del Perú, Fondo Editorial.

Harner, Michael. 1984. *The Jívaro: People of the Sacred Waterfalls.* Los Angeles: University of California Press.

ILO [International Labour Organization]. 1989. "C169—Indigenous and Tribal Peoples Convention, 1989 (No. 169)." Available at: http://www.ilo.org/dyn/normlex/en/f?p=NORMLEXPUB:12100:0::NO::P12100_ILO_CODE:C169 (accessed 9 January 2015).

Luna, Luis Eduardo. 2003. "Ayahuasca Shamanism Shared across Cultures." *Cultural Survival Quarterly* 27, no. 2: 20–23.

McMichael, Crystal H., Michael W. Palace, Mark B. Bush, Bobby H. Braswell, Stephen Hagen, Eduardo G. Neves, Miles R. Silman, Eduardo K. Tamanaha, and Chris Czarnecki. 2014. "Predicting Pre-Columbian Anthropogenic Soils in Amazonia." *Proceedings of the Royal Society B: Biological Sciences* 281, no. 1777: 2013–2475.

Mendoza Rodríguez and Rocío Elizabeth. 2007. *Irapay, Cosechando Hojas Hoy y Mañana.* Iquitos: Instituto de Investigaciones de la Amazonía Peruana, Proyecto BIODAMAZ.

MEA [Millennium Ecosystem Assessment]. 2005. "Linkages between Ecosystem Services and Human Well-Being." In *Ecosystems and Human Well-Being: Synthesis,* 88–102. Washington, D.C.: Island Press.

Mundo Shipibo. 2006. Available at: http://www.mundoshipibo.blogspot.com (accessed 31 July 2011).

Nikolakis, William and John Innes, eds. 2014. *Forests and Globalization: Challenges and Opportunities for Sustainable Development.* New York: Routledge.

Otsuki, Kei. 2013. *Sustainable Development in Amazonia: Paradise in the Making.* Routledge Studies in Sustainable Development. New York: EarthScan.

Rostain, Stéphen. 2012. *Islands in the Rainforest: Landscape Management in Pre-Columbian Amazonia,* vol. 4. Walnut Creek: Left Coast Press.

Rudel, Thomas K., Diane Bates, and Rafael Machinguiashi. 2002. "Ecologically Noble Amerindians? Cattle Ranching and Cash Cropping among Shuar and Colonists in Ecuador." *Latin American Research Review* 37, no. 1: 144–159.

Salisbury, David S. and Ben G. Weinstein. 2014. "Cultural Diversity in the Amazon Borderlands: Implications for Conservation and Development." *Journal of Borderlands Studies* 29, no. 2: 217–241.

Sarmiento, C. Alberto. 1961. *La Heroína de Motolo: Vida y Costumbres de los Jíbaros.* Comité Nacional Orientalista Ecuador Amazónico. Quito: Editorial La Unión.

———. 1970. *Embrujo Salvaje: Tradiciones y Cuentos de la Selva.* Quito: Editorial Fray Jodoco Ricke.

Vásquez, Rodolfo, Rocío Gonzáles, and Henk Van der Werff. 2010. *Flora del Río* Cenepa, Amazonas, Perú, vols. I–II. Missouri Botanical Garden Monographs in Systematic Botany, vol. 14. St. Louis: Missouri Botanical Garden Press.

Zimmerer, Karl S., ed. 2006. *Globalization and New Geographies of Conservation.* Chicago: University of Chicago Press.

Conclusion

Sarah Hitchner, Fausto Sarmiento, and John Schelhas

Amid increased recognition that exclusionary conservation is unjust and ineffective and that flawed models of "pristine wilderness" do not recognize anthropogenic landscape modifications, agroecological diversity, or the idea of sacredness imbued by humans on seemingly natural sites, there is a growing emphasis on incorporating land use history and sacred natural site (SNS) protection into landscape conservation planning. There is also increased awareness among conservation practitioners that documentation and analysis of cultural landscapes and sacred sites can simultaneously fulfill several goals: (1) promote a more detailed understanding of the landscape, (2) support indigenous rights, and (3) encourage collaboration between indigenous communities and conservation practitioners.

However, to date, many of these conservation efforts only superficially include indigenous and local community members—often only as holders of traditional ecological knowledge (TEK) and not as complex agents with rights to assess the pasts and guide the futures of these landscapes in which their cultures are embedded. At the same time, there have been numerous examples of successful conservation efforts involving SNS, driven both internally by indigenous and local communities and externally by organizations and agencies working collaboratively with them (Carmichael et al. 1994; Kelley and Francis 1994; Dudley et al. 2010; Vershuuren et al. 2010; Marchand et al. 2013). This volume has aimed to bring together examples and explanations of how the conservation of sacred sites throughout the Americas is contributing to the reinforcement, and in some cases revival, of indigenous identity and autonomy. As several of these terms are often at best misunderstood and at worst strategically misused, it is necessary here to explain our usage of these terms in this volume.

The term "indigenous" has been problematized by numerous scholars, and distinctions have been made between "indigenous," "native," "original," and other terms. In order for individuals and groups to be considered indigenous by both outsiders and insiders, certain terms and conditions must be met; these vary, of course, by country, region, and cultural milieu. For example, in the United States, there are often strict guidelines on who can claim membership in a Native American tribe based on ancestry and highly contested blood quantum laws, which calculate the fraction of a person's blood that is considered "Indian blood," as documented by either an official certification of Indian blood or a blood test that shows the genetic markers that prove native ancestry through DNA analysis (see Association on American Indian Affairs 2015). Karen Blu (1996: 224) writes about the external construction of Native American identity in relation to place:

> In the United States, American Indians are generally considered to be peoples with a special set of relations to federal and state governments. The issue of what a home place is and what sort of special connections indigenous peoples have to it has been and continues to be vital and repeatedly contested. Indian identities have seemed to hinge critically on their retention or reconstruction of and access to a home place, perhaps because displacement was so common and so devastating, so politically beyond the control of most groups throughout their histories. Whether one's group has a reservation or not is still a vital distinguishing feature among contemporary Native Americans.
>
> The reservation system is a product of the struggle between Indians and Whites for primacy of place—for the right to inhabit and control place and space. It has brought with it a peculiar fixing of the relations between Indian people and their reserved lands. In order to maintain legal claims to the land, Indians have had to present themselves as having been in the same place, often for time out of mind (or at least since the first European or American documents locate them somewhere), and as having had an unchanging kind of connection to their landscape. Anything that bolsters that notion—ancient myths recounting places of emergence, religious rites as old as any memories or written records, ancestors buried there—counts as evidence of "real" Indian existence to governmental agencies and the popular media. Such claims and demonstrations may be justified, but they also speak strongly to White notions that Indians who change are not "really" Indian and that entitlement to property is embedded in the legal system of this country.

Legal designation of a person as a Native American confers certain financial benefits—possibly including receipt of a portion of profits from casinos or extraction of natural resources such as oil in the West and the Arctic regions—and social and economic advantages that may result from scholarships, employment opportunities, and such that are reserved for Native Americans.

However, in addition to the obstacles faced by some individuals who self-identify as Indian or Native American but are not officially recognized as such, some groups of people that identify as a tribe are not recognized by the U.S. federal government. For example, the Lumbee of North Carolina in the United States are recognized as a state tribe, but they have been fighting for federal recognition since 1888; even though they are considered "Indians," they lack the benefits that federal recognition offers (Lumbee Tribe of North Carolina 2015). As Blu (1996: 201) states, "The Lumbee Indians have a 'homeland' that has never had even a semblance of protection by treaty or reservation. As far back as documentary evidence takes us, Lumbees have had to share this homeland with non-Native Americans." She also discusses the effect of outmigration from this homeplace on the Lumbee community and on Lumbee identity:

> For Indians, unless some children "stayed home" in the homeland, there would be no localized community, no focus, no peopled homeland. Lumbee identity, like the particular identities of other Native American peoples, depends heavily upon connection to a particular place, a home place. Lacking a home place, a home people, their identity would be a more generalized pan-Indian one, a free-floating kind of ethnicity with a very different emotional and conceptual construction. (Blu 1996: 207)

These examples of how the U.S. government recognizes some people and groups, and not others, as Indian or Native American demonstrate the complex interplay of historical, cultural, political, and physiological factors that converge to create indigeneity (which, as noted in the introduction, is a term used by some multinational entities in ways that include self-identification, not merely its imposition from outsiders). While in many cases indigeneity exists as a bureaucratic category created in alignment with states for purposes of governmentality, it is also fraught with deep cultural and social meanings as well.

The social and cultural complexities of this situation are too deep and convoluted to be disentangled and elucidated here, and we simply want to acknowledge this complexity and to explain how we conceptualize indigenous identity in this volume. As with Ross et al., we relate the term "indigenous" to the notion of identity; they write: "For us, 'Indigenous' is a category of identity, itself recognized by modern bureaucratic states, that emerges as much from current feelings of oppression and marginalization as from actual history" (2011: 24).

So here we do not consider the term "indigenous" to encapsulate an identity shared by a group of people and projected by them to outsiders. We acknowledge that, historically, indigenous groups have often been oppressed and marginalized by more powerful external entities (and many have also, at times, oppressed other groups). These groups are now strategically joining forces with other discrete indigenous groups to forge a common identity of "indigenous" and to strengthen their collective political power in a world that (we hope) is less hostile and more appreciative of indigenous contributions to nature and human society (see Witter et al. 2015).

In this volume, we recognize many documented cases of sacred rituals or religious beliefs as mechanisms that help safeguard natural resources (Callicott 1994; Berkes, Colding, and Folke 2000; Sponsel 2012; Tucker 2012; Anderson 2014), such as water (Lansing 1987, 1991, 2006; Ballestero 2012), game animals (Rappaport 1968; Balée 1985; Wadley and Colfer 2004), and marine resources (Johannes 1981; Acheson 1987, 2006). However, we take care to not romanticize indigenous knowledge or present it as a solution to ecological problems (see also Sillitoe 1998; Brosius 2000; Ross et al. 2011). As others have also noted, indigenous knowledge is not monolithic, static, or merely a consequence of collective memory and cultural transmission (Posey 1998; Berkes 2008). As with Western science, it is based on an accumulation of experiments, hypotheses, observations, syncretism, and adaptations (Berkes, Colding, and Folke 2000; González 2001; Tucker 2012). As Anderson (2014: 40–41) elegantly says:

> Negative judgements of traditional science are usually based on picking out the things that seem silliest to a modern Westerner, such as the idea that earthquakes are caused by a giant animal shaking himself below ground. Yet, such ideas are inferred causal mechanisms, similar to the above-mentioned phlogiston and aether. In *any* science it is these inferences that are most apt to be wrong and to be discarded over time. Many of Western science's inferred causal ideas, from string theory to the evils of saturated fats, are under attack today. This does not vitiate the accurate data assembled by modern science. Nor does the inferred underground animal's unreality vitiate the accuracy of traditional observations of where earthquakes occurred. It so happens that the underground animal, from China to Puget Sound, always shook himself in places where we now trace earthquake faults. These faults we now see as usually due to continental drift—an inferred cause that is now proven, but was ridiculed by almost all scientists in my college days.

Indigenous, or traditional, knowledge is experiential, shifting with new inputs of information and cultural influences, ranging from modern science and technology to globalized pop culture. New articulations of blended knowledge, often combined with strategic political and economic goals that require autonomy as a starting point, currently guide many indigenous groups to pursue not just autonomy but also the ability to become key players in the global market economy through various mechanisms, including niche products for export, ecotourism, and other methods of commodifying traditional products and knowledge. The concept that knowledge is inherently political, and that no form of knowledge is truly objective, has been explored by scholars representing many fields and disciplines (Foucault 1972, 1973, 1980; Said 1978; Escobar 1995; Agrawal 1995, 2002) and has been applied to many different conservation goals and initiatives (see, for example, Brush and Stabinksy 1996; Brush 1996; Brosius 1997, 2006; Sillitoe 1998; Nadasdy 1999, 2003; Brechin et al. 2002; Foale 2006).

We also do not want to limit indigenous knowledge to just TEK, acknowledging the resultant reductionist approaches to indigenous knowledge that fail to recognize the full complexity of knowledge relevant to many immediate and far-reaching issues of global, not just local, importance. Brosius and Hitchner (2010: 155) claim that while the valorization of indigenous knowledge has led to greater inclusion of such knowledge in conservation planning and natural resource management decision-making processes, there still remain power asymmetries in how such knowledge is applied and by whom. They write:

> That we are at last recognizing the value of local and indigenous knowledge rather than dismissing it as anecdotal or irrelevant is clearly a positive development. But by limiting our valorization of knowledge largely to that which pertains to the natural world, we subordinate such knowledge to the forms of knowledge possessed by decision-makers. Furthermore, one can draw a distinction between indigenous and local knowledge mediated by the research activities of social scientists and the knowledge articulated by local and indigenous activists and advocates themselves. One speaks in the passive voice of science—translating indigenous ways of knowing into forms intelligible to practitioners and decision-makers; the other speaks in the active voice of advocacy. Making this distinction draws our attention to the question of how local and indigenous perspectives and ways of knowing are elicited and translated between scales and how the link is made between this knowledge and the policy domain. (Brosius and Hitchner 2010: 155).

Indigenous knowledge is not just what is useful to modern science, and its value is not simply in its contribution to whatever lies at the intersection of different knowledge pathways; rather, it has intrinsic value, and the holders of it have the right to determine how it used and by whom.

The comanagement of indigenous territories has often been critiqued as being guided by Western science and continuing to exploit indigenous and marginalized groups by prioritizing conservation or preservation of natural areas over local groups' use of natural resources (Ross et al. 2011). Similar critiques of SNS conservation have been launched by skeptics of the idea that the protection of specific sites can preserve the functional integrity of full landscapes or watersheds and by critics who point out that protecting certain sites specific people deem sacred can lead to more contestation or to fragmentation of limited conservation resources.

Also, when sacred sites become recognized and popularized, they can become popular tourist destinations, with mixed results for local and indigenous communities. Schelhas notes that in addition to the ways that cultural tourism has the potential to "produce a commercialized stereotype of the culture, which can be degrading or constraining to local people," it can also lead to conflicts within and between indigenous communities. He notes examples in the United States of "the Havasupai and the Grand Canyon, the Navajo and Rainbow Bridge, and the Seminole and the Everglades" in which the sacred sites of local people become tourism destinations for outsiders (Schelhas 2002: 753–757).

We acknowledge these critiques of the term "sacred natural sites." We also accept SNS as just one category of protected area, one that can be overlaid with official designations—such as a United Nations Educational, Scientific and Cultural Organization (UNESCO) World Heritage Site or a National Historic Monument in the United States—understanding that it is not without complication and contention. It is not a universal category, and in some cases the governance of areas as SNS may prove to be overwhelmingly complex or physically impossible. Nevertheless, we contend that it is a useful category in many cases, one that articulates with other forms of conservation ranging from cultural landscapes to national monuments and historic buildings. We believe that bringing closer attention to the layers of history and focusing on the importance of place, in a world that is becoming increasingly interconnected and abstracted due to globalizing technologies and pop culture, is not only the most likely way to effect real change on the ground but also ethically the right thing to do.

But we also acknowledge that the use of the terms "sacred natural site" and "sacred landscape" leads naturally to several important questions, the answers to which are universal in some cases and very specific in most others:

1. What makes the landscape sacred?
2. To whom is the land sacred (now and in the past)?
3. What obligations do both native and nonnative people have to preserve the sacred nature of these sites and landscapes?
4. Who determines the outcome of the inevitable trade-offs that arise between preservation of these sites and developments that could cause irreparable physical or spiritual harm to these places?
5. Are there ways to maximize synergies between the goals of conservation and development in sacred areas?

None of these are new questions, and native and indigenous peoples, nonnative members of local communities, conservation and heritage protection practitioners, and scholars of disciplines ranging from archaeology to linguistics have grappled with these questions and their spinoffs for decades. It is no surprise that SNS are often located in highly contested areas, usually in areas rich with lucrative natural resources, and that the sacred nature of some natural places, including those with known human influence, is considered to be a benefit to conservation practices and goals. For example, Dove, Sajise, and Doolittle consider the case study of sacred groves, or sacred forests, and they cite examples of scholars who have documented varying levels of biodiversity conservation and deforestation rates in sacred forests. They note the attraction to Westerners of the idea of sacred forests, regardless of its actual impact on the ground:

> This image of non-Western peoples using spiritual doctrine to protect dwindling natural resources, often from Western-inspired or Western-driven development, has proven attractive to Western audiences (e.g., Chipko women tying themselves to trees in India or the Penan hunter–gatherers setting up blockades on logging roads in Borneo). Such practices seem to hold up a critical mirror to the secular character of resource-degrading life in industrialized nations. (2011: 7)

Thus, they argue, a narrow focus on the "sacred" aspects of conservation overlooks the practical and political reasons that indigenous and local communities protect specific areas and resources. However, as noted in the introduction to this volume, they also emphasize the strategic use of the notion of sacred spaces by indigenous and environmental activists. So although we acknowledge that the term "sacred," may be overstating, oversimplifying, or even misleading when applied to indigenous protection of certain areas, we also acknowledge its power to shape current conservation trajectories, and the contributors to this volume have explored this theme in various places throughout the Americas.

Another complication in the conservation of sacred sites is that sites considered sacred by one group are also often considered sacred by others, currently or in the past, leading to questions about whose sacrality should be prioritized in conservation efforts. Due to the frequent overlapping of religions and denominations in time and space, there is a long history of the sacred sites of one group being taken over and resacralized by the victors (as has happened with many Native American groups, for example, churches built on sacred sites). There is also a long history of syncretism, the blending of religious elements or spiritual explanations, conceptually, symbolically, and physically.

One example of institutionalized reinforcement of the dominant religion occurred when Catholic churches in the colonial city center of Quito, Ecuador, a city that was then being enlisted as a World Cultural Heritage Site, were restored, whereas indigenous sacred sites in and around Quito were not deemed worthy of significant UNESCO funding (Bromley and Jones 1995). It is also important to remember that the passing of time brings forth new generations, and even when cultural traditions and spiritual beliefs are transmitted more or less intact between generations, changes are inevitable, and these layers of history are evident in the landscape. Morphy noted:

> Landscape is a powerful factor in the operation of memory because of the associations narrators make between the local landscape and the events of the stories they tell. In these continual interactions between the past, the present, as well as the future, time is subordinate to place, and history as defined in the Western academic terms does not exist … Anthropologists have shown how ancestors and mythological events often become fixed in a specific landscape and act as timeless reference points. In a wider sense, individuals in all succeeding generations have to learn about the land, and in so doing, they not only renew the ancestral past, but they also transform it by adjusting it to present circumstances. (1995, qtd. in Christie 2009: xii)

These multiple layers of history and culture are evident in the present cultural landscapes of the Americas.

That sacred places in the Americas and around the globe are in danger is undisputed, and several recent events that resulted in damage to or destruction of sacred sites have been highly publicized. For example, in December 2014, prior to the conclusion of the 20th Conference of the Parties of the United Nations Framework Convention on Climate Change in Lima, Peru, some members of Greenpeace illegally entered the area around the famous Nazca (sometimes spelled "Nasca") Lines, large geoglyphs etched into thin desert crust. These activists placed big yellow letters on the ground to spell out: "Time for change! The future is renewable." This was done next to the geoglyph of a hummingbird, one of the largest and most recognizable of the figures.

The Nazca Lines have been on the UNESCO World Heritage List since 1994,[1] and access to the area around them is strictly prohibited. The president of Peru, Ollanta Humala, said the protesters showed a "lack of respect for our cultural patrimony and Peruvian laws" (Kozak 2014). Desecrating and irreparably damaging this SNS has publicized the discussion about using sacred sites for personal motives, with the symbolism of the place being an important motivator for both sides. Greenpeace, which has a long history of performing outrageous publicity stunts to call attention to issues they view as critical, lost a lot of support with this act, even among environmentalists who believe they literally crossed the line this time.

Many people who would ordinarily support and defend the rights of activists arrested during protests hope to see these activists penalized to the full extent of the law. Greenpeace has issued an official apology,[2] and Kumi Naidoo, Greenpeace's international executive director, has held meetings with officials in Peru to try to repair some of the social damage done by this act. However, as of the time of this writing, the Peruvian government has a pending lawsuit against the activists, and Luis Jaime

Castillo, Peru's vice-minister for cultural heritage at the time, said: "We are not ready to accept apologies from anybody ... Let them apologize after they repair the damage" (qtd. in Passary 2014).

Repairing the damage will likely prove to be impossible. Also, in February 2015, the first of the four major activists involved in the publicity stunt was arrested, after Greenpeace provided names to the Peruvian government (Ojeda 2015b); he has also publicly apologized for his role in the protest (Ojeda 2015a). As this drama continues to unfold, it becomes more apparent that sacred sites are in danger not just from developers, natural resource extraction companies, and other entities well known to cause ecological and cultural damage but also from groups and individuals usually considered allies in conservation and landscape preservation goals.

There are also other examples of the strategic or symbolic sacralization or desacralization of indigenous objects or ideas. For example, some indigenous people have emphasized the cultural roots of quinoa (*Chenopodium quinoa*) as the golden grain of the Andes and a symbol of indigenous identity, while it is simultaneously being marketed to organic food consumers of the Global North for the profit of private companies (Ruiz et al. 2014). Also, some ecotourism operations in the Amazonian rainforest have promoted the consumption of ayahuasca (*Banisteriopsis caapi*) to Western tourists seeking adventures and alternative healing experiences, which contrasts with the shamanistic ritualized uses of this plant (Tupper and Labate 2014). Also, controversy ensued following the polemical edict from Bolivian president Evo Morales to protect coca leaves (*Erythroxylum coca*) that recognized the sacred nature of the plant and the ancestral practices of chewing coca leaves in the Andes (Kim 2015).

That there remain threats and setbacks to conservation in general and the preservation of both tangible and intangible cultural heritage, and of sacred sites in particular, is hardly surprising. There have been many other documented cases of sites destroyed by various groups, including indigenous people themselves who have caused damage to their own sacred sites or to the sacred sites of other indigenous groups. However, both indigenous groups and the agencies and institutions that have historically mistreated them have led and approved a number of concerted efforts to protect indigenous heritage.

One key issue in protecting SNS is the process of reforming laws and policies so that these sites receive greater attention in the management of government protected areas, where many indigenous peoples' territories and sacred sites can be found today. There are no easy solutions to the challenge of getting government agencies—who have often taken into consideration what would be of most benefit to the general public in their management of nature conservation—to account for, respect, and appropriately manage sacred sites, but promising developments in the United States provide one example of a recent trend.

European colonization of what is now the United States was a process of continually disenfranchising Native Americans from their lands, with a significant amount of the lost land ending up in government protected areas (Schelhas 2002; Stevens 2014). Throughout history, protected areas were established on lands that had been traditionally occupied and used by Native Americans. In many cases, indigenous people were actively removed from established protected areas, and any indigenous rights granted in those protected areas at the time of establishment were often eroded through administrative action (Spence 1999; Schelhas 2002). Native Americans lost control of many of their sacred sites through these processes.

Although these actions came at a great cultural cost to native peoples and the search for remedies is often complicated, pathways are emerging that create opportunities to protect sacred sites on federal lands. Natural resource and protected area management worldwide has increasingly recognized and sought to incorporate the interests of indigenous people (Stevens 2014), and similar trends are emerging in the United States (Schelhas 2002).

In one example, the Forest Service, an agency of the U.S. Department of Agriculture (USDA), has recently focused considerable attention on sacred sites. Responding to a 2010 directive from the secretary of agriculture, a USDA team conducted more than fifty listening sessions with tribes and native peoples and reviewed existing laws, rules, regulations, and policies. A draft report was distributed to both tribal leaders and the general public for feedback, and the final report was released in 2012 (USDA 2012).

Importantly, the report recognizes that protecting and improving access to sacred sites are parts of the Forest Service's mission, and it further indicates support from the highest levels of government to seek a balance between the Forest Service's multiple use mandate, public needs and desires, and protection of sacred sites (USDA 2012: 10–11). Although clearly a compromise among sometimes competing interests, the report does encourage Forest Service employees to engage in dialogue and build relations with tribes and to be creative in seeking synergies and using existing laws and policies to protect sites (USDA 2012). Administrative pathways such as these, along with legal changes and activism, will be required to protect sacred sites on the ground.

Administrative, legislative, economic, environmental, and social incentives to protect SNS often intersect with the desire of indigenous groups to maintain, strengthen, or even revive their unique cultural identities.

This intersection can result in complementary efforts by different actors as well as in fractionation or contestation; the focus on protecting SNS can therefore result in what anthropologist Anna Tsing calls "friction," interactions during which cultures are coproduced through "awkward, unequal, unstable, and creative qualities of interconnection across difference" (2005: 4).

Although we have argued that there is a burgeoning solidarity among indigenous peoples around the world, often deployed strategically on larger national or international scales (see Witter et al. 2015), we do not wish to oversimplify the links between indigenous identity, cultural revival, biocultural diversity, and ecological conservation. We instead wish to emphasize that while these links do indeed exist and often reinforce each other, as is evident in many of the case studies presented in this volume, there is still much work to be done to build on these links and to fill in the persistent gaps between them and between them and all the forces currently threatening them. We hope that this volume is one small contribution toward that goal.

Sarah Hitchner, Ph.D., is an assistant research scientist at the Center for Integrative Conservation Research; an adjunct professor of anthropology at the University of Georgia, Athens; a cultural anthropologist; and an expert in sacred sites and cultural landscapes of Southeast Asia.

Fausto O. Sarmiento, Ph.D., is a professor of geography; the director of the Neotropical Montology Collaboratory, University of Georgia; and an expert in Andean cultural landscape conservation. Sarmiento was chair of the American Association of Geographers' Mountain Geography Specialty Group and the International Research and Scholarly Exchange Committee. He taught as a visiting professor in Costa Rica, Spain, Argentina, Chile, and Ecuador. He was awarded a plaque from the Centro Panamericano de Estudios e Investigaciones Geográficos ([CEPEIGE], the PanAmerican Center for Geographic Research and Studies). He is the author of *Montañas del Mundo: Una Prioridad Global con Perspectivas Latinoamericanas.*

John Schelhas, Ph.D., is a research forester at the Southern Research Station, USDA Forest Service, Athens, GA; a forester and anthropologist; and an expert in cultural aspects of forest and natural resource conservation and management in Latin America and the southern United States. He is the USDA Forest Service Southern Research Station copoint of contact for Tribal Activities, coordinating relationships between the Forest Service and Native American tribes in eleven southern states.

Notes

1. Greenpeace U.S. executive director Annie Leonard said, "The decision to engage in this activity shows a complete disregard for the culture of Peru and the importance of protecting sacred sites everywhere ... There is no apology sufficient enough to make up for this serious lack of judgment" (qtd. in Vergano 2014).
2. For example, UNESCO named the site "Lines and Geoglyphs of Nasca and Pampas de Jumana" (UNESCO 2007).

References

Acheson, James M. 1987. "The Lobster Fiefs Revisited: Economic and Ecological Effects of Territoriality in Maine Lobster Fishing." In *The Question of the Commons: The Culture and Ecology of Communal Resources,* edited by Bonnie J. McKay and James Acheson, 37–65. Tucson: University of Arizona Press.

Acheson, James M. 2006. "Institutional Failure in Resource Management." *Annual Review of Anthropology* 35: 117–134.

Agrawal, Arun. 1995. "Dismantling the Divide between Indigenous and Scientific Knowledge." *Development and Change* 26, no. 3: 413–439.

Agrawal, Arun. 2002. "Indigenous Knowledge and the Politics of Classification." *International Social Science Journal* 54, no. 3: 287–297.

Anderson, Eugene N. 2014. *Caring for Place: Ecology, Ideology, and Emotion in Traditional Landscape Management.* Walnut Creek, CA: Left Coast Press.

Association on American Indian Affairs. 2015. "Frequently Asked Questions." Available at: http://www.indian-affairs.org/resources/aaia_faqs.htm (accessed 8 February 2015).

Balée, William. 1985. "Ka'apor Ritual Hunting." *Human Ecology* 13, no. 4: 485–510.

Ballestero, Andrea. 2012. "The Productivity of Nonreligious Faith: Openness, Pessimism, and Water in Latin America." In *Nature, Science, and Religion: Intersections Shaping Society and the Environment,* edited by Catherine M. Tucker, 169–190. Santa Fe: School of American Research Press.

Berkes, Fikret, Johan Colding, and Carl Folke. 2000. "Rediscovery of Traditional Ecological Knowledge as Adaptive Management." *Ecological Applications* 10, no. 5: 1251–1262.

Blu, Karen I. 1996. '"Where Do You Stay At?': Homeplace and Community among the Lumbee." In *Senses of Place,* edited by Steven Feld and Keith H. Basso, 197–228. Santa Fe: School of American Research Press.

Brechin, Steven R., Peter R. Wilshusen, Crystal L. Fortwangler, and Patrick C. West. 2002. "Beyond the Square Wheel: Toward a More Comprehensive Understanding of Biodiversity Conservation as Social and Political Processes." *Society and Natural Resources* 15, no. 1: 41–64.

Bromley, Rosemary D. and Gareth A. Jones. 1995. "Conservation in Quito: Policies and Progress in the Historic Centre." *Third World Planning Review* 17, no. 1: 41–47.

Brosius, J. Peter. 1997. "Endangered People, Endangered Forest: Environmentalist Representations of Indigenous Knowledge." *Human Ecology* 25, no. 1: 47–69.

———. 2006. "What Counts as Local Knowledge in Global Environmental Assessments and Conventions?" In *Bridging Scales and Knowledge Systems: Concepts and Applications in Ecosystem Assessment,* edited by Walter V. Reid, Fikret Berkes, Thomas J. Wilbanks, and Doris Capistrano, 127–144. Washington, D.C.: Island Press.

Brosius, J. Peter and Sarah L. Hitchner. 2010. "Cultural Diversity and Conservation." *International Social Science Journal* 61, no. 199: 141–168.

Brush, Stephen B. 1996. "Whose Knowledge, Whose Genes, Whose Rights?" In *Valuing Local Knowledge Indigenous People and Intellectual Property Rights,* edited by Stephen Brush and Doreen Stabinsky, 1–24. Washington, D.C.: Island Press.

Brush, Stephen B. and Doreen Stabinsky, eds. 1996. *Valuing Local Knowledge: Indigenous People and Intellectual Property Rights.* Washington, D.C.: Island Press.

Callicott, J. Baird. 1994. *Earth's Insights: A Multicultural Survey of Ecological Ethics from the Mediterranean Basin to the Australian Outback.* Berkeley: University of California Press.

Carmichael, David, Jane Hubert, Brian Reeves, and Audhild Schanche, eds. 1994. *Sacred Sites, Sacred Places.* London: Routledge.

Christie, Jessica Joyce. 2009. "Introduction." In *Landscapes of Origin in the Americas: Creation Narratives Linking Ancient Places and Present Communities,* edited by Jessica Joyce Christie, x–xiii. Tuscaloosa: University of Alabama Press.

Dove, Michael R., Percy E. Sajise, and Amity A. Doolittle. 2011. "Introduction: Changing Ways of Thinking about the Relations between Society and Environment." In *Beyond the Sacred Forest: Complicating Conservation in Southeast Asia,* edited by Michael R. Dove, Percy E. Sajise, and Amity A. Doolittle, 1–34. Durham, NC: Duke University Press.

Dudley, Nigel, Shonil Bhagwat, Liza Higgins-Zogib, Barbara Lassen, Bas Vershurren, and Robert Wild. 2010. "Conservation of Biodiversity in Sacred Natural Sites in Asia and Africa: A Review of the Scientific Literature." In *Sacred Natural Sites: Conserving Nature and Culture,* edited by Bas Vershuuren, Robert Wild, Jeffrey A. McNeely, and Gonzalo Oviedo, 19–32. London: EarthScan.

Escobar, Arturo. 1995. *Encountering Development: The Making and Unmaking of the Third World.* Princeton: Princeton University Press.

Foale, Simon. 2006. "The Intersection of Scientific and Indigenous Knowledge in Coastal Melanesia: Implications for Contemporary Marine Resource Management." *International Social Science Journal* 58, no. 187: 129–137.

Foucault, Michel. 1972 [1969]. *The Archaeology of Knowledge.* Trans. Sheridan Smith. New York: Harper Colophon.

———. 1973. *The Order of Things.* New York: Vintage Books.

———. 1980. *Power/Knowledge: Selected Interviews and Other Writings, 1972–77.* Trans. C. Gordon. New York: Pantheon.

González, Roberto J. 2001. *Zapotec Science: Farming and Food in the Northern Sierra of Oaxaca.* Austin: University of Texas Press.

Johannes, Robert Earle. 1981. *Words of the Lagoon: Fishing and Marine Lore in the Palau District of Micronesia.* Berkeley: University of California Press.

Kelley, Klara and Harris Francis. 1994. *Navajo Sacred Places.* Bloomington: Indiana University Press.
Kim, Abraham. 2015. "The Plight of Bolivian Coca Leaves: Bolivia's Quest for Decriminalization in the Face of Inconsistent International Legislation." *Washington University Global Studies Law Review* 13, no. 3: 559–584.
Kozak, Robert. 2014. "Peru Says Greenpeace Permanently Damaged Nazca Lines." *Wall Street Journal,* 14 December 2014. Available at: http://www.wsj.com/articles/peru-says-greenpeace-permanently-damaged-nazca-lines-1418681478 (accessed 15 December 2014).
Lansing, Stephen J. 1987. "Balinese 'Water Temples' and the Management of Irrigation." *American Anthropologist* 89: 326–341.
———. 1991. *Priests and Programmers: Technologies of Power in the Engineered Landscape of Bali.* Princeton: Princeton University Press.
———. 2006. *Perfect Order: Recognizing Complexity in Bali.* Princeton: Princeton University Press.
Lumbee Tribe of North Carolina. 2015. "Lumbee Recognition." Available at: http://www.lumbeetribe.com/history--culture (accessed 14 December 2015).
Marchand, Michael E., Kristiina A. Vogt, Asep S. Suntana, Rodney Cawston, John C. Gordon Mia Siscawati, Daniel J. Vogt, John D. Tovey, Ragnhildur Sigurdardottir, and Patricia A. Roads. 2014. *The River of Life: Sustainable Practices of Native Americans and Indigenous Peoples.* Ecosystem Science Applications Series. East Lansing: Michigan State University Press.
Nadasdy, Paul. 1999. "The Politics of TEK: Power and the 'Integration' of Knowledge." *Arctic Anthropology* 36, no. 1–2: 1–18.
———. 2003. *Hunters and Bureaucrats: Power, Knowledge, and Aboriginal-State Relations in the Southwest Yukon.* Vancouver: UBC Press.
Ojeda, Hillary. 2015a. "Argentine Activist Apologizes for Greenpeace-Nazca Lines Incident." *Peru This Week,* 19 January 2015. Available at: http://www.peruthisweek.com/news-argentine-activist-apologizes-for-greenpeace-nazca-lines-incident-105031 (accessed 8 February 2015).
———. 2015b. "UPDATED: Nazca Lines: Argentine Police Arrest Greenpeace Activist." *Peru This Week,* 5 February 2015. Available at: http://www.peruthisweek.com/news-nasca-lines-argentine-police-arrest-greenpeace-activist-105206 (accessed 8 February 2015).
Passary, Anu. 2014. "What Are the Nazca Lines and Why Greenpeace Is Having a PR Nightmare." *Tech Times,* 14 December 2014. Available at: http://www.techtimes.com/articles/22133/20141214/what-are-the-nazca-lines-and-why-greenpeace-is-having-a-pr-nightmare.htm#ixzz3M1U9M9Is (accessed 15 December 2014).
Posey, Darrell A. 1998. "Comment on 'The Development of Indigenous Knowledge' by Paul Sillitoe." *Current Anthropology* 39, no. 2: 241–242.
Rappaport, Roy A. 1968. *Pigs for the Ancestors: Ritual in the Ecology of a New Guinea People.* New Haven: Yale University Press.
Ross, Anne, Kathleen Pickering Sherman, Jeffrey G. Snodgrass, Henry D. Delcore, and Richard Sherman. 2011. *Indigenous Peoples and the Collaborative Stewardship of Nature: Knowledge Binds and Institutional Conflicts.* Walnut Creek, CA: Left Coast Press.

Ruiz, Karina, Stephanie Biondi, Rómulo Oses, Ian Acuña-Rodríguez, Fabiana Antognoni, Enrique Martinez-Mosqueira, Amadou Coulibay, Alipio Canahua-Murillo, Milton Pinto, Andrés Zurita-Silvam, Didier Bazile, Sven-erik Jacobsen, and Marco Molina-Montenegro. 2014. "Quinoa Biodiversity and Sustainability for Food Security under Climate Change: A Review." *Agronomy for Sustainable Development* 34, no. 2: 349–359.

Said, Edward. 1978. *Orientalism.* New York: Pantheon.

Schelhas, John. 2002. "Race, Ethnicity, and Natural Resources in the United States: A Review." *Natural Resources Journal* 42, no. 4: 723–763.

Sillitoe, Paul. 1998. "The Development of Indigenous Knowledge: A New Applied Anthropology." *Current Anthropology* 39, no. 2: 223–252.

Spence, Mark. 1999. *Dispossessing the Wilderness: Indian Removal and the Making of the National Parks.* Oxford: Oxford University Press.

Sponsel, Leslie. 2012. *Spiritual Ecology: A Quiet Revolution.* Santa Barbara, CA: Praeger.

Stevens, Stan. 2014. "Indigenous Peoples, Biocultural Diversity, and Protected Areas." In *Indigenous Peoples, National Parks, and Protected Areas: A New Paradigm Linking Conservation, Culture, and Rights,* edited by Stan Stevens, 15–46. Tucson: University of Arizona Press.

Tsing, Anna Lowenhaupt. 2005. *Friction: An Ethnography of Global Connection.* Princeton: Princeton University Press.

Tucker, Catherine M., ed. 2012. *Nature, Science, and Religion: Intersections Shaping Society and the Environment.* Santa Fe: School of American Research Press.

Tupper, Kenneth W. and Beatriz C. Labate. 2014. "Ayahuasca, Psychedelic Studies and Health Sciences: The Politics of Knowledge and Inquiry into an Amazonian Plant Brew." *Current Drug Abuse Reviews* 7, no. 2: 71–80.

UNESCO [United Nations Educational, Scientific and Cultural Organization]. 2007. "Lines and Geoglyphs of Nasca and Palpa." Available at http://whc.unesco.org/en/list/700 (accessed 14 December 2016).

USDA [United States Department of Agriculture]. 2012. *USDA Policy and Procedures Review and Recommendations: Indian Sacred Sites. Final Report to the Secretary of Agriculture.* Washington, D.C.: USDA Forest Service Office of Tribal Relations.

Vergano, Dan. 2014. "Mystery Surrounds Delicate Nasca Lines Threatened by Greenpeace Stunt." *National Geographic,* 12 December 2014. Available at: http://news.nationalgeographic.com/news/2014/12/141212-nazca-lines-greenpeace-archaeology-science/ (accessed 15 December 2014).

Vershuuren, Bas, Robert Wild, Jeffrey A. McNeely, and Gonzalo Oviedo. 2010. *Sacred Natural Sites: Conserving Nature and Culture.* London: EarthScan.

Wadley, Reed L. and Carol J. Pierce Colfer. 2004. "Sacred Forest, Hunting, and Conservation in West Kalimantan, Indonesia." *Human Ecology* 32, no. 3: 313–338.

Witter, Rebecca, Kimberly R. Marion Suiseeya, Rebecca L. Gruby, Sarah Hitchner, Edward M. Maclin, Maggie Bourque, and J. Peter Brosius. 2015. "Moments of Influence in Global Environmental Governance." *Environmental Politics* 24, no. 6: 894–912.

Index

N